UNFROZEN GROUND: SOUTH AFRICA'S CONTESTED SPACES

For Muyahavho

Unfrozen Ground: South Africa's Contested Spaces

MAANO RAMUTSINDELA
University of the North, South Africa

Routledge
Taylor & Francis Group

LONDON AND NEW YORK

First published 2001 by Ashgate Publishing

Reissued 2018 by Routledge
2 Park Square, Milton Park, Abingdon, Oxon OX14 4RN
711 Third Avenue, New York, NY 10017, USA

Routledge is an imprint of the Taylor & Francis Group, an informa business

Notice:
Product or corporate names may be trademarks or registered trademarks, and are used only for identification and explanation without intent to infringe.

Publisher's Note
The publisher has gone to great lengths to ensure the quality of this reprint but points out that some imperfections in the original copies may be apparent.

Disclaimer
The publisher has made every effort to trace copyright holders and welcomes correspondence from those they have been unable to contact.

A Library of Congress record exists under LC control number: 2001088787

ISBN 13: 978-1-138-71174-7 (hbk)
ISBN 13: 978-1-138-71172-3 (pbk)
ISBN 13: 978-1-315-19957-3 (ebk)

Contents

List of Figures

List of Tables

Preface

Processes of transformation and reconstruction command growing attention in the analyses of post-1990 South Africa. Until recently, most analysts tended to explain these processes in general terms or paid attention to particular programmes in isolation. These approaches offer insights into pathways of post-apartheid transformation, but do not appropriately provide the much needed links between various threads of national reconstruction, and how these are contested on the ground.

This book attempts to provide those links by exploring the trajectories of the transformation of the state and society, and by relating nationally-driven processes to local articulations. Thus, the book attempts to simultaneously capture the dynamics of socio-political change at national and local levels.

On the whole, this book is biased towards themes that relate to my PhD project that was undertaken at Royal Holloway, University of London between 1996 and 1999. Out of that project emerged questions and issues that shaped the content and extent of this book. The first question concerns the abstraction of the post-apartheid state as an African state, hence the first part of the book provides a snapshot of experiences in Africa.

Against those experiences, the bulk of the book teases out emerging issues from abstract notions of identities and from experiences on land reform, and urban and rural changes. In this context, the second and major concern of this book is to understand strategies of addressing the legacy of apartheid and the attendant results.

In an attempt largely to explain post-1990 developments, I deal with complex issues that are very difficult to predict. I am, however, hopeful that confronting the legacy of apartheid head-on will take South Africa forward.

Maano Ramutsindela
Sovenga, 2001

Acknowledgements

There are many people to thank for this book. In retrospect, I do not think I would have thought about writing this book were it not for people who helped me to study for a PhD. I wish to thank Debby Potts who opened up possibilities for my studies in London. My sincere gratitude goes to David Simon who guided me throughout my research. Klaus Dodds commented helpfully on my work. The Department of Geography at Royal Holloway offered me the necessary academic environment and support.

I would like to thank Valerie Rose who suggested that I should think of writing this book. Following my book proposal, staff at Ashgate offered valuable assistance throughout the preparation of the book. The Assistant Desk Editor, Susan Hammant, was very helpful in the preparation of the final manuscript.

I am indebted to Canon Collins Educational Trust for Southern Africa (CCETSA) for the financial assistance. I also acknowledge the studentship from the Geography Department at Royal Holloway. The University of the North (South Africa) granted me study leave and gave me valuable support for which I am very grateful. My primary debt is to Takalani, Faresani and Thihangwi!

1 Introduction: Confronting the Past, Shaping the Future

Introduction: the colonial legacy

There is a tendency among commentators to describe dramatic political changes in hyperbole. In Africa the onset of decolonization in the 1950s and 1960s had been described as a wind of change, which Kwame Nkrumah (1961) of Ghana viewed as a raging hurricane that sweeps away the old colonialist Africa. The end of the Cold War in the 1980s was equally seen as unprecedented and momentous changes (Howell, 1994). Mason (1996, p.66) has in fact described the collapse of the Soviet Union in 1989 in these terms:

> This was a miraculous year in the history of post-war Europe. In the space of a few short months all the communist regimes in Eastern Europe within the Soviet sphere of influence had crumbled and disappeared.

These historical moments were, on the one hand, a time to celebrate political victories and freedom. However, on the other hand, they were periods in which the people of those countries were to confront serious problems. Such has been the case at the attainment of political independence in Africa in the 1960s, where socio-political challenges and problems of development were far too complex to grasp. In Central and Eastern Europe, too, political changes there were celebrated as the dawn of a 'new world order' free of the protracted Cold War, yet, the very change unleashed a wave of ethnic revivalist movements and brought economic turmoil in some parts of that region. More crucially, all these periods were followed by challenges of socio-political change and development.

In South Africa, the final collapse of the apartheid state in the 1990s signaled the end of the white minority domination and the beginning of a new era in which South Africans of all backgrounds could share the land of their birth. Those political changes were in part influenced by the crisis in civil-military relations in South Africa's Total Strategy; the end of

1

the Cold War that deprived the Total Strategy of any real intellectual credibility as the external communist threat to white power was removed; the failure of the strategy of inward industrialization (Mare, 1993; Cameron, 1995); and the intensity of national and international campaigns against apartheid. Thus, local and international factors combined to force the apartheid state to its knees.

Conversely, political changes in the early 1990s were followed by the momentous tasks of reconstructing the state and society: to de-racialize and restructure the state at all levels of government; to foster a new South Africanism; to address massive backlogs in education, housing, health, welfare, infrastructure, and so forth. All these were to be taken on board by post-apartheid governments. The transition from apartheid has achieved general currency in the analysis of post-1990 South(ern) Africa. On closer inspection, academics have variously captured either the general trends of that transition (see Marais, 1998) or the implementation of specific national programmes and/or strategies such as land reform (Levin and Weiner, 1997), regionalisation (Khosa and Muthien, 1998), nation-building (Maharaj, 1999) and the democratization of local government (Cameron, 1999). While these trends offer insights on challenges of transition in South Africa, they nevertheless neglect the connections between various national programmes and their impact on people on the ground.

Against that background, this book aims to make a contribution towards understanding the intricacies of national reconstruction in post-apartheid South Africa. It seeks to illuminate the conceptualization and implementation of national strategies that aim to address the legacy of apartheid, and to show how these are negotiated and/or contested on the ground. In pursuit of that aim, the volume attempts to simultaneously capture the dynamics of socio-political change at national, provincial and local levels. After all, apartheid was implemented at the micro-, meso- and national scales. Logically, any meaningful analysis of post-apartheid settings and challenges cannot be limited to any particular scale – the process of transformation cut across scales. It is a fairly worn argument that an appropriate scale of analysis and its associated boundaries is bound up with social process under investigation (Cox and Mair, 1988; Duncan and Savage, 1989). It is hoped that the attempts by this volume to capture the process of transformation within and across scales would not only enhance our understanding of that process in South Africa, but would also shed light on many and varied – old and (re) new(d) – forces that are at play. At the local level in particular, the beneficiaries of transformation are not and cannot be a passive recipient, while at the same time, most of those who were privileged by apartheid would like to 'hang on.'

Putting national reconstruction in gear

> In centuries of struggle against racial domination, South Africans of all colours and backgrounds proclaimed freedom and justice as their unquenchable aspiration. They pledged loyalty to a country which belongs to all who live in it. Those who sought their own freedom in the domination of others were doomed in time to ignominious failure ... Out of such experiences was born the vision of a free South Africa, of a nation united in diversity and working together to build a better life for all (Mandela, 10 December 1996).

These are the words of former President Mandela at the signing ceremony of the new constitution of the Republic of South Africa. Rightly so, the new constitution represents two historical moments: the ending of the dark past and the beginning of a new era of national reconstruction. It is common knowledge that South Africa's past has been largely shaped by years of colonialism and apartheid. The African National Congress (1994, p.3) aptly captured the effects of apartheid when it maintained that, 'there is not a single sector of South African society, nor a person living in South Africa, untouched by the ravages of apartheid.' Amin (1993, p.10) summed up the apartheid legacy in these words:

> Assembled on its territory [South Africa], one finds features proper to each and all of the 'four' worlds according to which countries can be classified. It has a white population which, in terms of habit and standard of living, belongs to the 'first' (advanced capitalist) world. A humourist might observe that the strong 'statist' behaviour of the white minority bears similarities to that of the former 'second' (so-called socialist) world. The black population of the townships clearly belongs to the modern industrializing 'third' world, and the 'tribal' peasants in the bantustans do not differ much from the peasant communities of what is now termed the 'fourth' world of Africa.

The precise number of 'worlds' within South Africa can be debated, and the list of the effects of apartheid is endless, but the deep scars of the legacy of that system are irrefutable. In the post-apartheid era, the crucial question has been, and still is, how to overcome that legacy and how to realize the vision of a non-racial democratic South Africa. Most notable, the nature and directions of post-1990 national reconstruction in South Africa were laid down in a negotiated political settlement that was brokered by both the

African National Congress (ANC) and National Party (NP) government as the main role players. Of significance in the negotiated political settlement was the emergence of a historic interim constitution in 1993. Whereas the interim constitution provided the template during the transition, the final constitution of 1996 sealed South Africa's road to democracy. As a supreme law of the country, the constitution became a useful instrument for the long-awaited democratic change.

While the vision of a non-racial democratic South Africa has been the hallmark of the struggle against apartheid, the realization of that vision hinges very much on post-apartheid national strategies and their outcomes. Such strategies and their underlying programmes were fiercely debated at the negotiation table and within political groupings. Various stakeholders sought to defend their long-held views, while at the same time, competing for agendas that could garner the political support of the populace and the goodwill of the international community. There had been competing demands for the form of the envisaged state; these ranged from the unitary state (ANC), federal state (NP and Inkatha Freedom Party) to secession (Afrikaner Volksfront and Freedom Front). Behind such demands lie the vexed question of the exercise of power and control over different sectors of the population and the use of that power in either the redistribution of resources or in defense of the status quo. To that end, the ANC that is 'historically prejudiced in favour of a strong and direct governmental role in the economy [a radical neo-socialist economic rhetoric of the 1970s and 1980s] has been forced to seize the nettle of state restructuring in line with local and global imperatives' (Munslow and FitzGerald, 1997, p. 44).

Subsequently, the ANC and alliance structures came up with the Reconstruction and Development Programme (RDP) in 1993, the aim being to use that programme as a tool for broader socio-economic transformation. Marais (1998, p.177) is of the view that, 'the RDP emerged as the most concerted attempt yet to devise a set of social, economic and political policies and practices that could transform South Africa into a more just and equal society.' Its basic principles of integration and sustainability, a people-driven process, peace and security, nation-building, reconstruction and development, and the democratization of the state (ANC, 1994) have significantly defined parameters for the agenda of transformation.

It must be noted that the RDP was a political party document and was employed as a manifesto of the ANC's election campaign in 1994. Nonetheless, the convincing victory of the ANC in the 1994 national election enabled that party to introduce the RDP as a national programme. The White Paper on the RDP committed all parties in the first non-racial

National Assembly to the objectives of that programme. Thus, the RDP became a unifying focus and symbol for the process of national reconciliation and socio-economic reconstruction (Blumenfeld, 1997; 1999). That programme has been criticized for its (over)emphasis on the objectives to be met with no parallel rigour on the analysis of mechanisms required for achieving those objectives. To that end, the macro-economic strategy, the Growth, Employment and Redistribution (GEAR) was introduced in 1996. It can be suggested that while GEAR remains largely focused on the goals of the RDP, it has embraced the market as an 'engine' of development. It has also profoundly shifted the ideals of the liberation movement from nationalization to privatization of state assets.

All in all, post-1994 national programmes are meant to address the effects of decades of apartheid rule and to build a united and unitary state. In a hitherto deeply divided society such as South Africa, such programmes are bound to affect both the material and non-material aspects of different communities. Attempts to address those divisions raise several questions: How could the vision of a non-racial society be achieved among people who had been deliberately divided on racial and linguistic lines? What happens when formerly divided communities are required to share the land, the cities and so forth? These questions are at the forefront of the various chapters of this book.

Scope and organization of the book

This volume is divided into two main parts. In the first part, Chapter 2, I locate the challenges of national reconstruction in South Africa within the post-independence/post-liberation problematic in Africa. The concern in Chapter 2 is to draw on - but not necessarily to compare - experiences in post-colonial Africa in order to illuminate challenges of transformation in South Africa. The analysis acknowledges the salient features of conditions in South Africa, but nonetheless endorses the view that South Africa has more in common with Africa than with Europe and North America (Simon, 1994; Mamdani, 1996; Simon and Ramutsindela, 2000).

The second part focuses on nationally driven processes that seek to achieve reconstruction and redress. Chapter 3 analyzes post-apartheid nation-building. At the abstract level, nation-building attempts to integrate 'citizens and subjects' through the drawing and crossing of identity boundaries. Here the notion of, and contestations over the 'rainbow nation' respectively reveal both the attempts to celebrate unity in diversity and the difficulties of transcending separate identities that have been nurtured over a long period of time. Given that the history of South Africa is broadly more the history of contested nationalisms (Mark and Trapido, 1987; Beinart, 1994), any meaningful transformation of society in that country has to confront issues of national identities.

Post-apartheid nation-building does not only occur at the abstract level(s), but is also reinforced by collapsing physical boundaries through land reform, provincial and local government boundary demarcations. Chapter 4 covers the land question that was central to the struggle against apartheid. It looks at the conceptualization and implementation of strategies to de-racialize patterns of land ownership and control. That general overview is followed by the analysis of the highly contentious Makuleke land claim in the northern eastern corner of the Kruger National Park. The Makuleke land claim was highly contentious as it involved more than 20 stakeholders, and drew statutory mining and environmental conservation policies into the ambit of land reform.

Chapter 5 reinforces the nature of contestations on the ground by looking at the re-mapping of towns and cities within the ambit of new provincial and local governments. Contestations over urban space are a reflection of those in society. The chapter also places the planner in the context of diverse conflicting interests. As Chapter . 6 illustrates, the contests over space are not limited to urban areas only, but are also found in rural areas where traditional structures of government are being reassessed. In Chapter 7 I draw the themes of the various chapters together and reflect on the implications of the route of transformation that South Africa has taken.

The focus on national reconstruction and the impacts at localities places severe limitations on this volume. On the one, it excludes global and regional dynamics that impinge upon pathways of transformation and development in post-apartheid South Africa. On the other hand, it is not possible to cover all national programmes in a single volume of this nature. Nevertheless, the impact of the selected national programmes such as nation-building, land reform and territorial restructuring on communities goes some way in illuminating obstacles and challenges of integrating a deeply divided society.

References

African National Congress. (1994), *The Reconstruction and Development Programme*, Umanyano, Johannesburg.

Amin, S. (1993), 'South Africa in the global economic system', *Work in Progress*, vol. 87, pp. 10–11.

Beinart, W. (1994), *Twentieth-Century South Africa*, Opus, Oxford.

Blumenfeld, J. (1997), 'From icon to scapegoat: South Africa's Reconstruction and Development Programme', *Development Policy Review*, vol. 15, pp.65–91.

Blumenfeld, J. (1999), 'The post-apartheid economy: achievements, problems and prospects', In J.E. Spence (ed.) *After Mandela: the 1999 South Africa elections*, Royal Institute of International Affairs, London.

Cameron, R. (1995), 'The history of devolution of powers to local authorities in South Africa: the shifting sands of the state control', *Local Government Studies*, vol.21, pp. 396–417.

Howell, J.M. (1994), Understanding Eastern Europe, Kogan Page, London.

Khosa, M. M. and Muthien, Y.G. (eds). (1998), *Regionalism in the New South Africa*, Ashgate, Aldershot.

Levin, R. and Weiner, D. (eds). (1997), *No More Tears: struggles for land in Mpumalanga, South Africa, Africa,* World Press, Trenton, NJ.

Maharaj, G. (ed). (1999), *Between Unity and Diversity: essays on nation-building in post-apartheid South Africa*, David Philip, Claremont.

Mamdani, M. (1996), *Citizen and Subject: contemporary Africa and the legacy of late colonialism*, Princetown University Press, Princetown.

Mandela, N.R. (1996), *Speech by President Mandela at the signing of the constitution*, Sharpeville, 10 December.

Marais, H. (1998), *South Africa – Limits to Change: the political economy of transformation*, Zed Books, London.

Mare, G. (1993), *Ethnicity and Politics in South Africa*, Zed Books, London.

Marks, S. and Trapido, S. (eds). (1987), *The Politics of Race, Class and Nationalism in Twentieth Century South Africa*, Longman, London.

Mason, J.W. (1996), *The Cold War: 1945-1991*, Routledge, London.

Munslow, B. and FitzGerald, P. (1997), 'The search for a development strategy: the RDP and beyond', In P. FitzGerald, A. McLennan and B. Munslow (eds), *Managing Sustainable Development in South Africa*, Oxford, Cape Town, pp. 41–61.

Nkrumah, K. (1961), *I Speak of Freedom*, Panaf Books, London.
Simon, D. (1994), 'Putting South Africa(n geography) back into Africa', *Area*, vol. 26, pp. 296–300.

Simon, D. and Ramutsindela, M.F. (2000), 'Political geographies of change in southern Africa', In R.C. Fox and K.M. Rowntree (eds) *The Geography of South Africa in a Changing World*, Oxford University Press, Cape Town, pp. 89–113.

2   South Africa 'in Africa'

Introduction

It seems naïve to use the phrase 'South Africa in Africa' while the country is in fact located on the continent. Nevertheless, behind that phrase lie some of the vexed questions that could be raised in the analysis of the South African state as a political domain. It being the outpost of a European polity on the continent, South Africa was not an African state. South Africa is not an exception in this regard, almost the whole continent constituted the colony of Europe. Furthermore, the sovereignty of colonial states was vested in Europe. The process of reverting that sovereignty back to Africa began through the struggle for liberation in the 1960s. In South Africa, the state remained an exclusive white polity till the early 1990s.

Against this background, the task in post-independence (South) Africa has been to reconstruct the state. That process was/is fraught with difficulties not least because of the colonial legacy and post-independence socio-political conditions. Generally, the record of state-building in Africa is not impressive, as we shall see below. That record is increasingly being employed by observers to frame questions of national reconstruction and state-building in post-apartheid South Africa. To the racists, the analytical framework seems simple: under black majority rule, the post-apartheid state will become just like other states in 'black' Africa. A more helpful framework would be to consider the legacy of apartheid and post-independence experiences on the continent, and to use that background in understanding the route and challenges of transformation in South Africa. This begs the question of why experiences in Africa are used to understand the reconstruction of the post-apartheid state. The reasons are not far-fetched: South Africa has more in common with Africa than with Europe and North America (Mamdani, 1996; Simon and Ramutsindela, 2000). For instance, it shares the similar bifurcation with other African states (Mamdani, 1996) and faces challenges of state-building that are typical of the post-independence era on the continent. These cannot be adequately understood by western theories of the state that are 'occupied with the reproduction of the modernist paradigms of the state and society, with what

9

Africa is not ...' (Berman, 1998, p. 306). There is a need to conceptualize African states in their own terms (Kamrava, 1993) without ignoring the general thrusts of theories of the state. Thus, there are salient features that differentiate African states from their European counterparts.

A double differentiation of the state in Africa

Many observers have problematized the state in Africa from different angles. The dominant approaches have been to look at Africa's colonial history and post-independence experiences. Colonial history has an upper hand because the contemporary state in Africa is largely shaped by the impact of, and its response to the colonization enterprise. It is important to note that Africa was not the only victim of colonialism, and was not even the first to be colonized (Asiwaju, 1996). Moreover, states on the continent are not the first to be integrated into the world economy on unequal and disadvantageous terms. The point here is that while Africa is different from the North, it shares common features with states in the South. For instance, Africa, Asia, the Pacific and Caribbean; all experienced the creation and imposition of sovereignty at the instance of the dominant states of the international system (Clapman, 1999). This is not to suggest that the South is a monolithic region. Rather, it is to appreciate the existence of the colonial legacy in a broader context.

Notwithstanding that colonial legacy, Africa has developed salient features as a result of its pre-colonial settings and the hybridization of those settings with colonial and post-colonial experiences – the 'grid of inheritance.' Colonialism has resulted in Africa becoming the most divided continent (Mazrui, 1980), with the most dependent position of all (Clapman, 1999). A growing body of literature draws a typology of peripheral states through the nature of state-society relationships. Kamrava (1993, p. 703) argued that, 'Third World polities can be conceptualized via a typology of state-society relationships according to which both states and societies assume a number of specific and mutually reinforcing characteristics.' In this context, Ergas (1987, p. 3) has emphatically argued that, 'African states ... have to a significant extent, a common matrimonial core.' According to Berman (1998), this patron-client relation has persisted in post-colonial societies and accounts for the personalistic, materialistic and opportunistic character of African politics. The general tenor of these arguments is that states in Africa could be differentiated from those in the South in general – the first differentiation. The salient features of African states do not necessarily mean that Africa is a homogenous region. Ergas

(1987, p. 3) is of the view that 'to reduce the multiple manifestations of African state to one, overall conceptual entity can be seen as presumptious, even perhaps arrogant, and justifiably so. After all, it can be argued that there are many African states as there are African countries.' Thus, while states in Africa share common features, those states are not necessarily the same – second differentiation. To that end, South Africa stands out as one of the most civil-minded state on the continent (Mamdani, 1996; Marais, 1998).

The double differentiation alluded to above begs the question of the extent to which the current nature of African states can be ascribed to a product of unique conditions and the impact of colonialism on the continent. For South Africa, the question could be asked: how does the specific character of South Africa opens the possibility of 'successful state-building'? For instance, does South Africa's progressive civil society and comparative high level of industrialization place the country in any different route to state-building than what has happened elsewhere on the continent. Insights into these questions can possibly be gained from the analysis of the process of national reconstruction. This chapter focuses on state-building in post-independence Africa. The process of reconstruction includes many elements that cannot be adequately handled in a single volume. The chapter pays particular attention to two aspects of reconstruction: nation-building and territorial reconfiguration. These aspects remain unresolved in most parts of the continent.

Confronting the colonial legacy

> For a truly comparative study of politics to develop, the great but incomplete drama of African state creation must be understood. This drama is as important to analyze as the process that led to the creation of France, Germany, and their neighbours. By examining both the environment that leaders had to confront and the institutions they created in light of their own political calculations, the entire trajectory of state creation in Africa can be recovered (Herbst, 2000, p. 30-31).

Few would dispute the complexities of state creation and reconstruction in Africa. In the view of the vastness of material on these complexities, I shall attempt to achieve brevity. Though Europeans employed their own model in the creation of states in Africa, states on the continent do not mirror their European counterparts. In this context, Young (1994, p. 283) is of the opinion that 'if we hold the colonial state up to the mirror of the "state", we find that the reflection is flawed.' The colony was denied sovereignty, the doctrine of the nation was disputed, the state was not an actor in the international scene, its territorial limits were highly artificial, and so forth. Sovereignty was problematic because the ultimate power was vested in the colonizing state. Existing African states were not recognized as states in international law and were therefore not entitled to exercise sovereignty over their territories (Fisch, 1988). That lack of recognition should not be construed as implying the absence of advanced polities at the time of European penetration. As Stock (1995) has shown, each of the major regions of the continent had advanced kingdoms and empires before colonial conquest. These polities lost their identity in the colonial melting pot (Wallace-Bruce, 1985). Besides problems of sovereignty, Africans were not citizens but subjects in the polity in which they had been enclosed (Mamdani, 1996). Young (1994, p. 43) observed that, 'the doctrine of the nation, redolent with overtones of self-determination, was vigorously disputed by the propriety powers until the eve of their departure.'

None of the colonial powers intended to foster national cohesion among the indigenous people. Instead, they preserved and exacerbated stratification. Ranger (1983, p. 249) argued that 'the immobilization of populations, re-enforcement of ethnicity and greater rigidity of social definition in 20th century Africa are a result of both the necessary and unplanned consequences of colonial economic and political change, ... and of the conscious determination on the part of the colonial authorities to "re-establish" order and security and a sense of community by means of defining and enforcing "tradition."' While ethnic differences were exacerbated in Africa (Coleman, 1994), it was discouraged in Europe (Short, 1993). That is, nation-building in Africa was not in the interest of metropolitan Europe. Even during the Cold War, neither superpower supported nation-building on the continent and elsewhere. Mayall (1992, p. 21) contends that, 'until 1991, there was no indication that the breakup of existing states would be countered by the international community.' In Africa states were split asunder by forces aligned to the superpowers.

The nature of nationhood in the colony also affected the position of that polity in the international scene. The international law at the time governed inter-state relations as those between civilized nations. As Africa was not accepted into the 'family of civilized nations', states there could not become actors in the international scene.

All these flaws were to be faced by the post-independence African leaders. Sovereignty was regained through negotiations and armed struggles, and sometimes a combination of both. In the case of former colonial masters, negotiations were seen as a platform on which liability of colonial states could be resolved. For instance, the end of slavery meant that Ghana could become a liability to Britain (Boateng, 1978). In the same vein, French colonies could vote whether to retain colonial links with France. This is not to suggest that Europe willingly surrendered sovereignty to Africa – in most cases it was compelled to do so. This is clearer in the armed struggle that became a factor in the liberation of Madagascar, Algeria, Angola, Guinea-Bissau, Mozambique, Zimbabwe, Namibia and South Africa. The armed struggle in these countries played an important role in forcing the white oligarchy to surrender political power.

It is important to note that the process of decolonization not only reverted sovereignty to the African people, but set in motion the arduous challenge of consolidating power over inherited territories. Against this background, Herbst (2000, p. 11) asserts that 'the fundamental problem facing state-builders in Africa – be they pre-colonial kings, colonial governors, or presidents of the independent era – has been to project authority over inhospitable territories that contain relatively low densities of people.' Nonetheless, demography alone cannot account for challenges of state-building on the continent. What needs to be asked is how post-independence leaders responded to challenges of state-building on the continent.

State-building and builders

> Efforts to build states in Africa take place in a setting of two dimensions: in time, as part of a history of indigenous culture and political traditions, colonial domination, nationalist independence movements, and the trials of new nationhood; and in place, as the activities of members of a world system of integrated economies and political relations, with a well-defined and sometimes brutally administered pecking order of power and status (Stark, 1986, p. 335).

It is a truism that state-building in Africa should be understood from the vantage point of the history of the continent and should also be viewed in the wider context of the inter-state system. There is therefore a need to understand historical moments in the construction of states on the continent. According to Doornbos, (1990, p.181) 'the time of independence is not necessarily the most appropriate time from which to begin an analysis of state formation and performance because structural determinants inherited from the colonial era set definitive limits to the actions of the state and to a large extent predetermine the trajectories of its formation.' Such structural determinants are evident in inherited territorial limits within which nationhood was to be forged to complete the statehood of post-independence African polities. The section below pays particular attention to responses to these determinants.

The territorial question

It is inconceivable to think of a state without any geographic attachment. Indeed, the classical formulation of the criteria for statehood requires a state to have a defined territory. That is, the definition of state must give weight to territorial limits over which sovereignty is exercised. Africa inherited the problem of territorial limits from colonialism.

In Africa, the territorial limits of states are problematic, mainly because boundaries there bear little or no resemblance to pre-colonial polities. As a gamut of literature shows, the territorial limits in present day Africa are a reflection of European interests on the continent. Europeans curved the map of Africa to establish their political domains. However, the process by which the map was drawn has generated much heated academic debate. Central to the debate is the part played by the Berlin Conference of 1884-5 in the partitioning of the continent (see Hargreaves, 1988; Katzellenbogen, 1996). At issue is whether Africa was partitioned at the Conference. Notwithstanding that debate, what is relevant to the discussion of this chapter is the existence of territorial limits that are at variance with the historical geography of the continent – lumping together different population groups while splitting same communities into different countries. Of course, such tendencies are not unique to Africa. However, a case can be made that the mismatch between territorial limits and population groups is most excessive in Africa than elsewhere.

Though the problem of territorial limits is ascribed to colonialism, it was perpetuated in the post-colonial era, the more so because the post-colonial state was erected artificially on the foundations of the colonial state (Chabal, 1986; Bayart, 1993). That colonial legacy means that the territorial question is bound up with any efforts towards reconstructing the state on the continent. To date that question has not been successfully addressed because it is feared that territorial reorganization would open Pandora's box; a view held by the Organization of African Unity (OAU). The perpetuation of the fragmentation is also ascribed to the acceptance of the nation-state in international politics, the effects of the Cold War, the determination of the political elite to hold onto colonial heritage, and lack of interest on the part of 'partitioned Africans' in shifting boundaries (Ramutsindela, 2000). All these mean that the territorial question in Africa has been kept on hold. Herbst (2000, p. 25) argues that,

> far from being a hindrance to state consolidation, African boundaries have been perhaps the critical foundation upon which leaders have built their states. In addition, the territorial boundaries help shape other buffer institutions that also insulate polities from international pressures.

Viewed from this angle, the territorial arrangement on the continent offered politicians the opportunity to manipulate the colonial legacy for self-interest. As Mazrui and Tidy (1984) have noted, African leaders are to a large measure to blame for the perpetuation of the balkanization of the continent. However, it is worth noting that few African leaders recognized the danger of maintaining colonial arrangements. For instance, Nkrumah (cited in New African, 2000, p. 21) warned that, 'unless we succeed in arresting the danger [of territorial fragmentation] through mutual understanding on fundamental issues and through African unity, which will render existing boundaries obsolete and superflous, we shall have fought in vain.' In paying homage to Nkrumah, Pan-Africanist scholars have argued that unless boundaries are changed, decolonization remains an unfinished business. To that end, Soyinka (cited in Asiwaju, 1998) has suggested that we should sit down with square-rule and compass and re-design the boundaries of African nations. Generally, views on how to deal with the problem of artificial boundaries fall into three broad categories. The first of these is to redraw the political map of Africa into larger and fewer states as suggested by Gakwandi (1996) and Makua (cited in Asiwaju, 1998).

The second position is to redraw boundaries in order to create smaller states. Bello (1995, p. 546) has argued that 'what we must now do is re-group peoples and re-draw boundaries on a rational and logical basis to take cognisance of linguistic, cultural and ethnic diversities as is the case with most successful nations all over the world.' Bello's view will, according to Nyerere (*New African*, January 2000, p. 31), fossilize Africa into a worse state than the one it is in. The third view advocates the use of regional blocks as an avenue through which meanings of colonial boundaries could be changed without necessarily redrawing boundaries (Asiwaju, 1998; Ramutsindela, 2000). As Asiwaju (1996) has argued, existing boundaries could be turned into theaters of opportunities by emphasizing their integrative process. African leaders can perhaps be forgiven for not changing external boundaries, but there could be little sympathy for their failure to change internal territorial arrangements that had been instrumental to colonial rule. It is worth noting that, internal structures such as regions, districts and 'tribal areas' were created to assist in the administration of the colonial state. How that was done will become clear in the analysis of the South African state in subsequent chapters. Nevertheless, it is worth noting that post-independence African leaders placed emphasis on the unity of their countries on the one hand, while embracing divisive territorial arrangements on the other (Sidaway, 1992; Simon and Sidaway, 1993, Ake, 1996). That contradiction is more visible in the institution of local government, as internal divisions are mostly bound up with that institution. In the post-independence era, conservative states such as Lesotho and Botswana incorporated local structures ('traditional councils') into the apparatus of the state. In few cases such as Tanzania, traditional authorities were effectively abolished after independence. As we shall see below, the restructuring of local government in South Africa was strongly contested between 1993 and 2000.

One can argue that more insights into the problems of boundaries on the continent could be gained from a broader conception of the notion of a boundary. Recent conceptualizations of boundaries are anchored on the social and symbolic construction between social collectives rather than viewing boundaries as mere physical limits of states. Boundaries constitute social and cultural practices which are sources of meanings and interpretations (Paasi, 1996). These broader conceptualizations are important for devising solutions to boundary disputes. To that end, most of the suggested solutions to Africa's territorial problems are state-centric. Agnew (1999, p. 175) has described a state-centric perspective as 'thinking and acting as if the world were made up entirely of states governing blocks of space which between them constitute the politico-geographical form of world politics'. This territorial trap is more evident in strategies that (over)emphasize the redrawing of the map, as I have shown above.

It must be emphasized that post-independence African leaders failed to provide a new territorial framework in which the post-colony could be constructed. They were more concerned with seizing the sovereignty of the state than with changing the territorial limits over which that sovereignty was to be exercised. As Clapman (1999, p. 523) puts it, 'the newly independent governments ... took to Westphalian sovereignty like ducklings to water.' The problem of inherited territorial limits is closely related to that of building nations within defined boundaries. A closer inspection of nation-building on the continent reveal the complexity of the process.

Forging nations in the absence of a national space

Discussions on the post-colonial African state have centred largely on state power and capacity and on national identity and unity. Attention was and is being paid to issues of national identity because of the association of a nation with the state, and because the national question on the continent is a thorny issue. The association of nations with states reflects a political principle that seeks the congruence between the political and the national unit (Gellner, 1983). The principle itself is sanctioned by international law by which 'people are given their identity by the state within which they find themselves' (Knight, 1985, p. 249).

Colonial design and the demand for nation-states by international law meant that post-independence African leaders were to fit nations into existing territorial arrangements. Since colonial boundaries had arbitrarily divided same communities and lumped different ones together, nation-building on the continent had been fraught with difficulties. In the run up to independence, challenges of nation-building were mostly obscured by solidarity against the former colonizers as various groups identified more closely with the centre (Doornbos, 1990, p. 193). That is, anti-colonial wars enabled liberation movements to 'enlarge local and ancestral loyalties, often as deeply felt as they were divisive, into a unifying vision' (Davidson, 1981, p. 139). In the post-independence era, leaders considered nation-building as necessary for keeping the state together – fear of disintegration (Nyerere, 1968). The leaders were indeed to be seen by the international community as speaking on behalf of their nations. Thus, the new leadership was required to satisfy the demand of modern statehood, i.e. nationhood.

It would seem the post-independence era was characterized by lack of a unifying element of nationhood. Herbst (2000) is of the view that the new states could not exert authority over a sparsely populated terrain. He goes on to say that, 'relatively low population densities in Africa have automatically meant that it has always been more expensive for states to exert control over a given number of people compared to Europe and other densely settled areas' (Herbst, 2000, p. 11). While the demographic factor might help us understand the problem of exercising authority by states on the continent, it falls short of explaining the complex problem of nation-building. The concept of nationhood demanded a redefinition of communities into a nationhood that was not necessarily compatible with how communities viewed themselves and their ways of life. As Doornbos and Markakis (1994) have shown in the case of Somalia, pastoralists there did not have states nor did they need one, hence, the state was a mismatch did not have states nor did they need one, hence, the state was a mismatch between state and society. Such a mismatch is aptly captured in the notion of the societal disengagement in which sectors of society withdraws from the state (Miles, 1995). An argument can be made that the definitions of nationhood in the post-independence era are based on European concepts of a nation. As Mazrui and Tidy (1984, p. 85) have cautioned, 'typically nineteenth-century European definitions of nationality based on European political norms of nations with centralized governments and fixed boundaries do not and need not apply to Africa. Politically decentralized nations consisting of ethnic groups which lacked political unity and central political institutions were often united by a common language and culture.'

The definition of nationhood on the continent is therefore loaded with ideological questions. Europeans believed that Africans belong to 'tribes', despite the existence of groups of people who are no less than a nation. Lonsdale (1992, p. 19) argued that the highland African such as the Kamba, the Kikuyu, Embu and Meru 'were peoples, not tribes, potential nations rather than actual dispersions of related lineages.'

Notwithstanding the state-society divide, leaders of the post-colony were less successful in transcending the divide among communities that had been enclosed within common boundaries. Historically, such division was nurtured by colonialism. Nonetheless, a divided nationhood in post-independence societies was underscored by various factors. These include the struggle by different population groups to control the resources of the state. That struggle is often associated with the absence of a public realism at the face of the plural societies. As Hyden (cited in Carroll and Carroll, 1997, p. 478) has noted, 'the trend of post-independence politics in most African countries has been to disintegrate the civic public realm inherited from the colonial powers and replace it with rivaling communal or primordial realms, all following their informal rules.' One can argue that the problem is not so much the disintegration of the civic realm, but rather the alienation of that realm from the African masses. It should be noted that indigenous people did not participate in the civic realm of the colony (Mamdani, 1996). They were supposed to do so after independence. In the majority of cases, the few who had access to the public realm exploited it for sectarian purposes. The source of disintegration may well include the existence of multiple realms in tandem. As Membe (1992, p. 5) has put it, the postcolony is made up not of one coherent public space, nor is it determined by any single organizing principle. It is rather a plurality of spheres and arenas, each having its own separate logic yet nonetheless liable to be entangled with other logics when operating in certain specific contexts '… the postcolonial subject mobilizes not just a single identity, but several fluid identities which, by their very nature, must be constantly revised in order to achieve maximum instrumentality and efficacy as and when required.'

This begs the question of whether the failure to forge a common nationhood in the post-colony could be ascribed to the plurality of society there. The question become even more critical if we acknowledge that plurality is not a unique African phenomenon. What needs to be accounted for is why the multiple identities continue to overshadow a shared nationhood. The answer lie partly in looking at the process of nation-building itself. Literature shows that that process has tension between models of nationhood. The nature of the inherited state required a western-oriented nationhood – after all, the inherited state was western! Unsurprisingly, post-colonial leaders had to adopt that notion of nationhood. However, the process of adoption was also problematic. In the quest for national unity, most leaders paid scant attention to forms of identities that were later to threaten the envisaged unity. Furthermore, that quest was not matched by serious search for bases on which the nationhood could be imagined. Generally, nationalism has been mobilized from raw materials such as culture, history and other inheritances of the pre-nationalist world. Evidence of this is found in the use of Gaelic in the construction of Irish nationalism (Johnson, 1992).

In most African states, such raw materials were not mobilized. As Gellner (1983, p. 81) has noted, 'the nationalism of sub-Saharan Africa often neither perpetuate nor invent a local high culture, nor do they elevate an erstwhile folk culture into a new, politically sanctioned literate culture, as European nationalism has often done.' Davidson (1990, p. 14) is of the view that, 'nationalism [in Africa] went wrong because the nationalists turned their backs on what they said was traditional Africa ... They turned their backs on everything that African history had so far achieved, in terms of self-development.'

Attempts to invoke African history were made by Pan-Africanist leaders such as Nkrumah, who cherished the one-ness of Africans. However, such ideals sailed against the glorification of the nation-state in the post-independence era. In few cases such as Tanzania where the Swahili (language) was mobilized, the nation-building project yielded some positive results. All in all, nation-building in Africa failed at both the continental and national levels. The record of state-building in Africa raises questions of how that process will proceed in post-apartheid South Africa (Ramutsindela, 2001). While more than half the states on the continent have been involved in boundary disputes, the post-1990 South Africa had to resolve its dispute with Namibia over the Orange River boundary, and over the Walvis Bay enclave (see Hangula, 1993; Simon, 1996). The solutions to the dispute were successfully negotiated in 1993 and 1994 respectively. However, South Africa, like its counterparts on the continent, has to face the problems of nation-building and the re-arrangement of its internal spaces in line with the vision of a non-racial society. Chapter 3 below shows the trajectories and contestations over nation-building.

References

Agnew, J. (1999), 'The new geopolitics of power', In D. Massey, J. Allen and P. Sarre (eds), *Human Geography Today*, Polity Press, Cambridge, pp. 173–193.

Ake, C. (1996), *Democracy and Development in Africa*, Bookings Institution, Washington, D.C.

Asiwaju, A.I. (1996), 'Borderlands in Africa: a comparative research perspective with particular reference to Western Europe', In P. Nugent and A.I. Asiwaju (eds), *African Boundaries: barriers, conduits and opportunities*, Pinter, London, pp. 253–65.

Asiwaju, A.I. (1998), 'Fragmentation or integration: what future for African boundaries?' Paper presented at the Fifth International Conference of the Boundaries Research Unit on 'Borderlands Under Stress', 15–17 July 1998, Durham.

Bayart, J-F, (1993), *The State in Africa: the politics of the belly*, Longman, London.

Bello, A. (1995), 'The boundaries must change', *West Africa*, p. 546.

Berman, B.J. (1998), 'Ethnicity, patronage and the African state: the politics of uncivil nationalism', *African Affairs*, vol. 91, pp. 305–41.

Boateng, E.A. (1978), *A Political Geography of Africa*, Cambridge University Press, Cambridge.

Carroll, B.W. and Carroll, T. (1997), 'State and ethnicity in Botswana and Mauritius: a democratic route to development', *Journal of Development Studies*, vol. 33, pp. 464–486.

Chabal, P (ed.). (1986), Political Domination in Africa: reflections on the limit of power, Cambridge University Press, New York.

Clapman, C. (1999), 'Sovereignty and the Third World state', *Political Studies*, vol. 47, pp. 522–537.

Coleman, J.S. (1994), *Nationalism and Development in Africa*, University of California, Berkeley.

Davidson, B. (1981), *The People's Cause: a history of guerrillas in Africa*, Longman, London.

Davidson, B. (1990), 'The crisis of the nation-state in Africa', (interviewed by Munslow), *Review of African Political Economy*, vol. 49, pp. 9–21.

Doornbos, M. (1990), 'The African state in academic debate: retrospect and prospect', *Journal of Modern African Studies*, vol. 28, pp. 179–198.

Doornbos, M. and Markakis, J. (1994), 'Society and State in Crisis: what went wrong in Somalia'? *Review of African Political Economy*, vol. 59, pp. 82–89.

Ergas, Z. (1987), 'Introduction', In Z. Ergas (ed.), *The African State in Transition*, Macmillan, London, pp.1–22.

Fisch, J. (1988), 'Africa as terra nullius: the Berlin Conference and International Law', in S. Forster, W.J. Mommsen and R. Robinson (eds), *Bismarck, Europe and Africa: The Berlin Africa Conference 1884-1885 and the onset of partition*, Oxford University Press, Oxford, pp. 347–375.

Gakwandi, A.S. (1996), 'Towards a new political map of Africa' in T.Abdul-Raheem (ed.), *Pan Africanism: politics, economy and social change in the twenty-first century*, Pluto, London, pp. 181–190.

Gellner, E. (1983), *Nations and Nationalism*, Blackwell, Oxford.

Hangula, L. (1993). *The International Boundary of Namibia*. Gamsberg-Macmillan, Windhoek.

Hargreaves, J.D. (1988), 'The Berlin Conference, West African boundaries and the eventual partition', in S. Forster, W.J. Mommsen and R. Robinson (eds) *Bismarch, Europe and Africa: The Berlin Africa Conference 1884-1885 and the onset of partition*, Oxford University Press pp. 313–20.

Herbst, J. (2000), States and Power in Africa: comparative lessons in authority and control, Princeton University Press, Princeton, N.J.

Johnson, N.C. (1992), 'Nation-building, language and education: the geography of teacher recruitment in Ireland, 1925-55', *Political Geography*, vol. 11, pp. 170–189.

Kamrava, M. (1993), 'Conceptualising Third World politics: the state-society see-saw', *Third World Quarterly*, vol. 14, pp. 703–716.

Katzellenbogen, S. (1996), 'It didn't happen at Berlin: politics, economics and ignorance in the setting of Africa's colonial boundaries', In P. Nugent and A.I. Asiwaju (eds), *African Boundaries: barriers, conduits and opportunities*, Pinter, London, pp. 21–34.

Knight, D.B. (1985), Territory and people or people and territory? Thoughts on postcolonial self-determination. *International Political Science Review*, vol. 6, pp. 48–272.

Lonsdale, J. (1992), 'The conquest state of Kenya 1895-1905', In B. Berman and J. Lonsdale, *Unhappy Valley (Book Two)*, James Currey, London, pp. 13–44.

Mamdani, M. (1996), *Subject and Citizen: contemporary Africa and the legacy of late colonialism*, Princeton University Press, Princeton, N.J.

Marais, H. (1998), South Africa – Limits to Change: the political economy of transformation, Zed Books, London.

Mayall, J. (1992), 'Nationalism and international security after the Cold War', *Survival*, pp. 19–35.

Mazrui, A. (1980), *The African Condition: a political diagnosis*, Heinemann, London.

Mazrui, A. and Tidy. M. (1984), *Nationalism and the new states in Africa*, Heinemann, London.

Mbembe, A. (1992), 'Provisional Notes on the Postcolony', *Africa*, vol. 62, pp. 3–37.

Miles, W.F.S. (1995), 'Decolonization and disintegration: the disestablishment of the state in Chad', *Journal of Asian and African Studies*, vol. 30, pp. 41–52.

New African. (2000), January, no. 381.

Nyerere, M.J. (1968), *Nyerere: Freedom and Socialism*, Oxford University Press, Nairobi.

Paasi, A. (1996), Territories, Boundaries and Consciousness: the changing geographies of the Finish-Russian border, Wiley, New York.

Ramutsindela, M.F. (2000), 'African boundaries and their interpreters', In N. Kliot and D. Newman (eds), Geopolitics at the end of the twentieth century: the changing world political map, Frank Cass, London, pp. 180–198.

Ramutsindela, M.F. (2001), 'Down the post-colonial road: reconstructing the post-apartheid state in South Africa', *Political Geography*, vol. 20, pp. 57-84.

Ranger, T. (1983), 'The invention of tradition in colonial Africa', In E. Hobsbawn and T. Ranger (eds), *The Invention of Tradition*, Cambridge University Press, Cambridge, pp. 211–262.

Sidaway, J.D. (1992), Territorial organization and spatial policy in post-independence Mozambique in historical and comparative perspectives (PhD Thesis), University of London, London.

Simon, D. (1996), 'Strategic territory and territorial strategy: the geopolitics of the Walvis Bay's reintegration into Namibia', *Political Geography*, vol. 15, pp. 193–210.

Simon, D. and Sidaway, J. (1993), 'Geopolitical transition and state formation: the changing political geographies of Angola, Mozambique and Namibia', *Journal of Southern African Studies*, vol. 19, pp. 6–28.

Simon, D. and Ramutsindela, M.F. (2000), 'Political geographies of change in southern Africa', In R.C. Fox and K.M. Rowntree (eds), *The Geography of South Africa in Changing World*, Oxford University Press, Cape Town, pp. 89–113.

Short, J.R. (1993), *An introduction to political geography*, Routledge, London.

Stark, F.M. (1986) 'Theories of contemporary state formation in Africa: a reassessment', *Journal of Modern African Studies*, vol. 24, pp. 335–347.

Stock, R. (1995), *Africa South of the Sahara: a geographical interpretation*, Guildford, New York.

Wallace-Bruce, N.L. (1985), 'Africa and International Law – the emergence of statehood', *Journal of Modern African Studies*, vol. 23, pp. 575–602.

Young, C. (1994), *The African Colonial State in Comparative Perspective*, Yale University Press, New Haven.

3 Nation-building in South Africa

Introduction: crossing non-spatial boundaries

According to Simpson (1994), any discussion on nation-building in South Africa would take place against the pessimistic scenario of the inability of the state in Africa to create a central focus for the loyalties of their citizens who are seen to be trapped in their 'pre-modern' parochialisms. The underlying view is that nation-building, as a major thrust in post-colonial Africa has failed – so why would it succeed in South Africa? Such a view not only oversimplifies the diverse outcomes of nation-building, but also attempts to write the future of South Africa in advance.

In terms of the theme of this book, this chapter aims to show that South Africa, like the other post-colonial states on the continent, has to address the question of national identity. As I have argued in Chapter 2, the problems of national identities in post-independence Africa can be fully understood when the nature of the colonial state is taken into consideration. To that end, the first part of this chapter attempts to locate the politics of nationalism(s) in South Africa in its historical context. In that part, I show how successive governments manipulated the identities of the different population groups. That discussion is followed by the second part that analyses mechanisms and strategies, that aim to deal with the legacies of fractured identities. The main focus of the second part is the process of, and debate over the (re)construction and (re)crossing of identity boundaries.

The politics of nationalism(s) in South Africa

It is perhaps not an overstatement to suggest that the history of South Africa is broadly more the history of contested nationalisms. Any discussion of national identity in South Africa has to pay attention to the salient meanings of identity constructs. Dubow (1994, p. 356) cautioned that, in South Africa, the term ethnic has acquired particular sorts of meanings and associations that are heavily context-dependent. Ethnic is, for example, variously synonymous with words like population groups, tribe, nation, *volk* and race.

This chapter employs the terms 'population groups', 'race' and 'ethnic groups' separately in order to aid understanding of the various models and interpretation of nation-building in South Africa. My use of these terms neither imply acceptance of their legitimacy as deployed by the apartheid regime, nor the existence of fixed identities. The terms sand categories are only helpful in understanding the historical contexts and meanings of identity labels existing at the time.

The meanings attached to categories of people in South Africa are more than semantics, because they tend to reflect constructions of the 'other'. For instance, the Afrikaners considered themselves as a *Volk* (with capital V) but described black people as *volkies* (with small letter v) in order to separate a superior 'nation' from inferior 'tribes' (February, 1991). In the same vein, the English contemptuously considered the Afrikaner as an inferior 'race', thus, creating the ambiguity in the relationship between settlers themselves. Despite their differences, British and Dutch settlers considered all Africans in South Africa as inferior as shown by the following remarks:

> The white man must rule because he is elevated by many, many steps above the black man [i.e. African]... which it will take the latter centuries to climb and which it is quite possible that the vast bulk of the black population may never be able to climb at all (Lord Milner, cited in Marks and Trapido, 1987, p.7).

> As against the European, the native stands as an eight-year old against a man of mature experience ... Differences exist in ethnic nature, ethnic custom, ethnic development and civilization and all these differences shall long exist (Hertzog, cited in Marks and Trapido, 1987, p. 9).

These racial stereotypes suffused into the making of a white nation in 1910 when the Union of South Africa was formed. The Union constitution that was drafted after the all-white National Convention of 1908 codified white supremacy (De Kiewiet, 1941; Karis and Carter, 1972; Manzo, 1992). However, the white nation was short-lived as Afrikaner nationalists redefined themselves as a distinct nation different from the English (Giliomee, 1989, 1992).

The construction of Afrikaner national identity had a profound impact on African nationalism. As Marks and Trapido (1987, p. 1) have shown,

> for much of the twentieth century, an exclusive form of white Afrikaner nationalism, with its explicit objective the capture of the state by the white Afrikaner 'nation', has confronted its counterpart, a pan-South African black nationalism, which has sought the incorporation of Africans into the body politic.

Thus, the exclusion of Africans from the Union bolstered the platform on which African nationalism could be mobilized beyond the simple 'tribal' identities (Benson, 1985). This kind of mobilization is reminiscent of liberation movements elsewhere on the continent, as I have shown in Chapter 2. Marks (1986, p. 54) argued that it is 'in this pan-South African political arena [that] the black elite saw their way forward through an inclusive, liberal-democratic nationalism'. As shown below, such a pan-South Africanism was rejected by successive white minority governments, and also created fertile grounds for opposition from a section of African nationalists.

As shown above, Afrikaners redefined their own identity within a white nation. However, such a redefinition was not accompanied by an Afrikaner homeland. Arguably, the 87 per cent of the land that was set aside for white occupation and ownership (see Chapter 4) amounted to a territorial home for Afrikaners and the English. If the English and the Afrikaners were distinct 'races' as Afrikaner ideologues wanted us to believe, separate territories would have been a logical step. Instead, only Africans were assigned to a separate 13 per cent of the land (see Chapter 4). Thus, the construction of 'race' was conveniently and inconsistently employed to divide the land between black and white people, while at the same time, being employed as a tool to mobilize Afrikaner identity.

Furthermore, while Afrikaner identity was mobilized through, *inter alia*, the use of language, the so-called coloured who speak Afrikaans were excluded mainly because they were considered a 'mixed race', or 'the illegitimate progeny of European civilization and non-European savagery' (see Marks and Trapido, 1987, p. 29). The manipulation of language in identity construction is not surprising because language is often singled out as an essential criterion in the making of nationhood. Smith (1976, p. 18) has argued that, only through a vernacular and its literature can one grasp the 'personality' of a nation; and only through communication of messages in an agreed code can a sense of nationality develop.

Unlike the Africans, coloureds were at least regarded as having 'European blood'; a racist ideology that made coloureds hierarchically superior to Africans. This hierarchical re-ordering of identities was translated into government policies, one of which was the Coloured Labour Preference Policy. This policy was meant to limit the employment and residence of Africans in the Western Cape and to give preference to the employment of coloured people. Although coloured identity cannot be attributed to state policies alone, policies of preference had strong influence in the making of such an identity. According to Goldin (1987, p. 27),

> coloured identity reflected in part the determination of the skilled stratum of the coloureds to defend their position vis-à-vis the African population and to provide a means to assert their claim for preferential treatment.

In this context, the policies of preference served to differentiate the coloureds from both the Africans and the whites, thus placing coloureds as a distinct intermediate group. Goldin (1987, p. 27) went further to argue that,

> the continued existence of an intermediate [coloured] group ... depended on the success of policies which sought to promote the interests of coloured people relative to Africans whilst at the same time preventing the assimilation of coloured and white people.

The racial definitions set in motion by Afrikaner nationalists not only created contradictions within Afrikanerdom, but also created conditions for the rise of what Shula Marks (1986) called, 'the ambiguity of nationalism' among African nationalists. For their part, African nationalists had to reconcile the ideals of a South Africanism founded on European values and the realities of the hitherto organization of African societies. Coleman (1994) has noted that African nationalists on the continent were influenced by European civilization. The same can be said about elements of African nationalism in South Africa.

More generally, a perception of the need for Africans to unite against white domination existed and nurtured the aspiration for a pan-South Africanism (Lodge, 1983; Beinart, 1994). This vision is succinct from the resolution taken at the All-African Convention (AAC) in 1935. The AAC resolved to,

> ensure the ultimate creation of a *South African nation* in which, while various racial groups may develop on their own lines socially and culturally, they will be bound together by the pursuit of common political objectives (Walshe, 1970, p. 121).

This vision of a non-racial South Africa was enshrined in the Freedom Charter of 1955, and remained central in the thinking of the ANC (Meli, 1988). The Charter envisaged a society in which the white and non-white peoples of the Union will work and live together in harmony for the common good of the fatherland. A counter view was that African people can be organized only under the banner of African nationalism as shown by the founder of the Pan Africanist Congress (PAC), Mangaliso Sobukwe (1959, p. 20):

> We aim, politically, at the government of the Africans by the Africans for Africans, with everybody who owes his only loyalty to Africa and who is prepared to accept the democratic rule of an African majority being regarded as an African. We guarantee no minority rights, because we think in terms of individuals, not groups.

The versions of nationhood articulated by the ANC and PAC were not acceptable to Afrikaner nationalists because neither of them served Afrikaner interests. Instead, successive Afrikaner governments sought to keep Africans and whites as very much separated neighbours but not fellow citizens while, at the same time, promoting separate national units among Africans. Such a separation culminated in the creation of bantustans in which ethnically homogenous groups would fuse into separate 'African nations'.

For their part, Indian immigrants in South Africa had to construct their identities under different circumstances. Indian immigrants did not have a common identity in South Africa due to class and religious divisions (Marks and Trapido, 1987; Swan, 1987). For this reason, Swan (1987, p. 183) has argued that,

> the only reason, therefore, that one can speak of 'Indian' politics or 'Indian community' in South Africa as early as 1890 is the fact that both colonial and republican states in South Africa classified all emigrants

from the sub-continent - and their descendants, even if South African born - as Indians.

Nonetheless, the claim to a South Africanism by Indians helped to foster solidarity between Indians and Africans because both groups were being denied citizenship by the apartheid state (Ramsamy, 1996). The leader of the Transvaal Indian Congress, Yusuf Dadoo, captured the sense in inter-group solidarity in these words: 'we have entered into a period of active co-operation between the oppressed peoples for basic human rights' (cited in Swan, 1987, p. 203). Notwithstanding this solidarity, the prejudices between Africans and Indians re-surfaced in different forms under different circumstances. Material conditions emerged as an important factor in the construction of identities in apartheid South Africa.

Material conditions and identity construction

Since the opening of the mining industry in South Africa in the nineteenth century, the English dominated the economy of the country. Terreblanche and Nattrass (1990) consider the Union period between 1910 and 1922 as characterized by the economic and political dominance of the English establishment. This dominance resonated with a superior English identity, a position which Afrikaner nationalists aimed to match (Adam and Moodley, 1993). In this context, the English-dominated capitalist development gave rise to Afrikaner economic nationalism, a movement that O'Meara (1983) has referred to as *volkskapitalisme* (people's capitalism). The economic movement had to deal with two important issues. First, most of the Afrikaners were farmers and were not trained in commerce and industry. The revenue of the farmers was therefore the main resource that could be used to drive *volkskapitalisme*. Financial institutions such as the *Suid Afrikaanse Nationale Trust & Assuransie Maatskappy* (Santam) and *Suid Afrikaanse Lewens Assuransie Maatskappy* (Sanlam) were established in order to 'pool the money resources of all Afrikaans-speakers in a central fund, there to be converted into productive capital' (O'Meara, 1983, p. 99). The motive of Sanlam was clearly spelt out in the chairman's report of 1921:

> Sanlam is an authentic institution of the Afrikaner *volk* in the widest sense of the word. As an Afrikaner you will naturally give preference to an Afrikaner institution. I would just remind policy holders that we are busy

furnishing employment to young Afrikaners, and training them in the assurance field (cited in O'Meara, 1983, p. 98).

Second, the English dominated capitalist economy disrupted the way of life of different communities in South Africa. It, for instance, created conditions for employment in the urban sector that led to the migration of blacks and whites. The migration of cheap labour from white farms caused a shortage of labour on the farms. Those farmers who migrated to urban areas had to compete for jobs with their black counterparts. Furthermore, over 80 per cent of poor whites in the country in the 1930s were Afrikaners (O'Meara, 1983). These conditions were taken into consideration in the mobilization of Afrikaner identity. The poor white (Afrikaner) problem featured strongly in the *Ekonomiese Volkskongres* (Economic Congress of the volk) of 1939, as the remarks by one of the main speakers suggest:

> So long as nearly 300,000 Afrikaans-speakers live below the bread-line; so long as a percentage of our fellow Afrikaners remain the hewers of wood and drawers of water in their country; so long as the Afrikaner is notable by his absence in our business life; and so long as the large section of the agrarian population are forced by circumstances to migrate to the cities in order to make a living, millions of pounds belonging to Afrikaners lie around unproductively (cited in O'Meara, 1983, p. 113).

This congress, which O'Meara (1993), Marks and Trapido (1987) and Terreblanche and Nattrass (1990) consider to be one of the great turning points in the development of Afrikaner nationalism, helped to mobilize Afrikaners towards a political victory that saw the National Party taking over the state in 1948. From that date, programmes to uplift the Afrikaner were pursued with much rigour. As Terreblanche and Nattrass (1990) observed, the National Party enlarged the bureaucracy and parastatal sector in order to generate Afrikaner employment opportunities.

A corollary to the rise of Afrikaner identity and hegemony was the impoverishment and suffering of the Africans, coloureds and Indians. The identities of these groups were officially defined by the Population Registration and the Group Areas Acts of 1950. These acts aimed to classify the population groups, and the components of urban areas in which they could live and own property respectively. For their part, Africans were mainly confined to their respective bantustans and urban 'native locations', a political vision that led to massive forced removals and resettlement schemes – with incalculable costs in terms of human suffering in the process. The institutionalization of Afrikaner dominance in the 1960s was facilitated by the economic boom that financed social engineering by pouring vast sums of money into the bantustans. However, these political structures were not politically and financially viable.

They remained vulnerable even in the face of the tricameral parliament that was underwritten by the 1983 constitution. The aim of the constitution was, among other things, to protect the self-determination of population groups (South Africa, 1983; Munro, 1995). As a result of this constitution, separate Houses of Parliament were established for whites, coloureds and Indians. The inclusion of coloureds and Indians into, and the exclusion of Africans from, the organs of the state in part reflect the racial hierarchies and stereotypes that had been nurtured by successive white minority governments. Effectively, the inclusion of coloureds and Indians into national governance did not imply that the boundaries of white identity had collapsed. The arrangement only meant that coloureds and Indians could become part of 'white' South Africa but not part of a white nation (February, 1991).

The 1980s were also a period of intensive resistance by black and progressive white people. That combination of opponents of apartheid had the mirror image of a South African nation of many 'colours' – the rainbow nation – hence Archbishop Desmond Tutu (1995, p. 183) could declare that, 'this country [South Africa] is a rainbow country ... We say we are the rainbow people. We are the new people of the new South Africa.' The ideals of a rainbow nation were pursued after 1990.

Reconstructing a rainbow nation

Difficulties in creating a shared nationhood in South Africa since 1910 has led analysts to conclude that South Africa was a state without a nation. Unsurprisingly, the national question featured in the first round of negotiations at the Convention for a Democratic South Africa (CODESA) in 1991, where the majority of political parties declared an undivided South Africa with *one nation*. However, conservative Afrikaners and bantustans leaders in the former Ciskei, Bophuthatswana and KwaZulu rejected the declaration in favour of a federal dispensation. These reduced the new South Africanism to the protection of groups whose integration is seen to be incompatible with the primordial nature of the population of that country, i.e., neo-apartheid. In the 1990s as in the 1960s, conservative Afrikaners seek to protect their identity in isolation in the form of a *volkstaat* (Ramutsindela, 1998). Strategies for achieving that goal range from a military coup to securing certain areas for Afrikaner domination in local government.

For its part, the NP moved away from group identities based on statutory classification (De Klerk, 1990), seeking to protect white identity within a federal system in which minority groups would have control over their own affairs – a continuation of the won affairs mentality. In this context, the NP proposed a regional demarcation that encapsulates both federalism and the respect for regionally based identities.

The ANC, however, remains committed to its long-held view that South Africa belongs to all who live in it, black and white. Its view is that South Africa should be transformed into a non-racial society in which race and ethnicity cannot be used as building blocks for a new South Africanism. It propagates the individual's rights and freedom as essential elements of the new South Africanism – as articulated in the Bill of Rights. Marks (1995, p. 2) has argued that 'at common sense level, the non-racialism propagated by the ANC is the demand that an individual's citizenship, legal rights, economic entitlements and life-chances should not be decided on the basis of racial ascription.' The non-racialism espoused in the Freedom Charter has been adopted as a preamble to the new constitution (ANC, 1994).

This begs the question of how a non-racial society can be built in a country characterized by very different material conditions. As the then Deputy and now President Mbeki (1998a) pointed out,

> a major component part of the issue of reconciliation and nation building is defined by and derives from the material conditions in our society which have divided our country into two nations, the one black and the other white. We therefore ... say that South Africa is a country of two nations.

In the same vein, material conditions of black people are symbolic of the 'oppressed black nation'. As the ANC intellectual and cabinet minister at the time, Pallo Jordan (1997, p. 4) has put it,

> National oppression thus found expression in the palpable form of a number of economic, social and developmental indicators – such as poverty and underdevelopment, the low levels of literacy and numeracy among the oppressed communities, their low access to clean water, the non-availability of electricity, their low food consumption, their invariably low incomes, the poor state of their health, their low skills, the generally unsafe environment in which these communities lived, etc.

It follows that socio-economic conditions cannot be ignored in the process of nation-building in present-day South Africa. Thus, the basic needs of the black majority remain fundamental to post-apartheid nation-building processes. Thus, the process of building a rainbow nation is bound to affect or to be affected by the historically constructed hierarchies of privileges between the whites, Africans, coloureds and Indians. In other words, socio-economic conditions in South Africa run parallel to the question of race, hence attempts to address the legacy of apartheid are bound to impinge upon the race question. Indeed, the central objective of the RDP is 'to improve the quality of life of all South Africans, and in particular the most poor and marginalized sections of our communities' (ANC, 1994, p. 15).

However, the new government's intentions to improve the lives of the black people are misconstrued by white-dominated political parties (i.e. the DP and NNP) as shunning away from the ideals of a rainbow nation. For instance, the DP (1998) argues that it is immoral and impractical to try to redress racial imbalances by taking steps that will entrench race consciousness. Instead, it offers a vision of a South Africa consisting of an opportunity society. It can be suggested that nation-building in South Africa should include strategies to overcome the racial divide. The problem of race and nation-building is more pronounced in South Africa than elsewhere on the continent because of the large number of whites in the country. Mamdani (1996, p. 28) observed that the 'sheer numerical weight of white settler presence in South Africa sets it apart from settler minorities elsewhere in colonial Africa.' However, that distinction presents more difficulties than solutions to state-building.

In recent months, debate on how to overcome the racial divide intensified, culminating in the national conference on racism on 30 August 2000. The conference was convened to facilitate public discussion on strategies to overcome racial divisions. There are two main divergent views on how the problem might be resolved. First, there is a view that discussions on racism might encourage racial conflict, threaten national reconciliation and prevent the birth of a rainbow nation. A point can be made that racism cannot disappear on its own. This leads to the second view that the dream of a non-racial society cannot be achieved if attention is not paid to the manifestations and effects of racism, all the more because the divisions of the South African society were anchored on a racist ideology. As President Mbeki (2000) reminds us, 'the social and economic structure of our society is such that the distribution of wealth, income, poverty, diseases, land, skills, occupations, intellectual resources and opportunities for personal advancement, as well as the patterns of human settlement, are determined by the criteria of race and colour.' Racial divisions cannot be wished away!

Nation-building in South Africa not only faces the challenge of racial divisions, but also has to engage with ethnic divisions among the black people. It should be noted that bantustans were 'cynically created to manipulate impulses towards nationhood, based upon the invention of traditional and cultural identity, whilst, importantly, at the same time, overriding emphasis was upon the need for the transformation and modernization of these peripheral backwaters' (Jones, 1999, p. 582). The net effects of the bantustans on the identities of Africans who had been pushed to those areas need to be assessed. For the moment, the government has given official recognition to African languages, and supports the revival of African cultures. However, the distribution of resources to these groups will play a significant role in shaping the outcome of nation-building in South Africa.

Conclusion

In this chapter I have attempted to show that the question of national identities in post-apartheid South Africa could not have been ignored, because the history of South Africa was basically underlined by that of contested nationalism. As shown in the analysis, the construction of white (Afrikaner) identity and the appropriation of the state as a white polity, heightened the assertion of identities by the black communities. Successive white minority governments sought to foster fractured identities through strategies and for reasons I have discussed above.

The chapter also shows that the construction and deconstruction of national identities were and continue to be contested. Unlike national struggles that seek to purify space, that in South Africa aimed at sharing the territory of the state by those who live in it. It is against this background that the approach to nation-building in the post-apartheid era has been to seek to achieve unity in diversity, hence the notion of the rainbow nation. That notion, though, faces many challenges, including the recent demand for a *volkstaat* for white minority groups. Perhaps the main challenge would be the new government's ability to convince members of the different population groups (and their sub-groups) to pay allegiance to a broader South African nationalism. President Mbeki, then Deputy President, (1998b) captured this challenge in this way:

We must therefore pose the question to ourselves as to whether the diversity of which we speak, with seeming pride, is a blessing or a curse; whether the fact of a heterogeneous society makes for an easier achievement of the goal of a better life for all, or whether we would have been better served if we had a more homogenous society ... If the real problem we face, of ending the legacy of the past persists ... it will not be because we are cursed with the gift of diversity ... [it would be] because we would have failed to find ... common national aspirations and a common identity.

Challenges to nation-building are not unique to South Africa because nation-building, as Gellner (1983) has shown, has never been a smooth process. Even though the nation-building is affected by ideological movement, national solidarity, the degree of internal politico-cultural unity, collective suffering, or a combination of sets of factors, the results of that process have never been the same. There is therefore no 'better way' towards nation-building. The success of that process is contingent on approaches to nation-building, and the dynamics of centrifugal and centripetal forces, and how these are managed. For the moment, the overwhelming majority of South Africans identify with South Africa in territorial terms. Thus, there is a widespread identification with South Africa as a territory. This, according to Guelke (1996), is a reflection of the presence of some society-wide loyalties. Against the backdrop of such 'territorial nationalism', Simpson (1994, p. 473) concluded that, 'a new democratic government [in South Africa] will start with a distinct advantage over many states in Asia and Africa, where it is precisely the territorial state that has often been contested.'

However, it would be wrong to assume that the apparent patriotism has put the national identity problem to rest, state-driven processes such as land reform (Chapter 4), the redemarcation of provincial and local government boundaries (Chapters 5), continue to impinge upon racial and ethnic consciousness. All these processes are, as we shall see, necessary in reconstructing the post-apartheid state. The process of land reform to which I now turn illustrates this point.

References

Adam, H. and Moodley, K. (1993), *The opening of the apartheid mind*, University of California Press, Berkeley.

African National Congress. (1994), *The Reconstruction and Development Programme*, Umanyano, Johannesburg.

Beinart, W. (1994), *Twentieth-century South Africa*, Opus, London.

Benson, M. (1985), *South Africa: the struggle for a birthright*, International Defence and Aid Fund for Southern Africa, London.

De Kiewiet, C.W. (1941) *A history of South Africa: social and economic*, Oxford University Press, London.

De Klerk, F.W. (1990), *Address by State President F.W. De Klerk at the opening of the parliament of the Republic of South Africa*, 2 February, Cape Town.

Democratic Party. (1998), *The death of the rainbow nation: unmasking the ANC's programme of re-racialization, Policy Document*, Johannesburg.

Dubow, S. (1994), 'Ethnic euphemisms and racial echoes', *Journal of Southern African Studies*, vol. 30, pp. 355–70.

February, V. (1991), *The Afrikaners of South Africa*, Kegan Paul, London.

Gellner, E. (1983), *Nations and nationalism*, Blackwell, Oxford.

Giliomee, H. (1989), 'The beginnings of Afrikaner ethnic consciousness, 1850-1915', In L.Vail (ed.), *The creation of tribalism in southern Africa*, University of California Press, Berkeley pp. 21-54.

Giliomee, H. (1992), 'Broedertwis: intra-Afrikaner conflicts in the transition from apartheid', *African Affairs*, vol. 91, pp. 339-364.

Goldin, I. (1987), *Making race: the politics of economics and coloured identity in South Africa*, Longman, London.

Guelke, A. (1996), 'Dissecting the South African miracle: African parallels', *Nationalism and Ethnic Politics*, vol. 2, pp. 141-54.

Jones, P.S. (1999), '"To come together for progress": modernization and nation-building in South Africa's bantustan periphery – the case of Bophuthatswana', *Journal of Southern African Studies*, vol. 25, 579–605.

Jordan, Z.P. (1997), *The national question in post 1994 South Africa, Discussion paper for the ANC's 50th National Congress*, Mafikeng.

Karis, T. and Carter, G.M. (1972), *From protest to challenge: a documentary history of politics in South Africa, 1884-1964, vol.1*, Hoover Press, Stanford.

Lodge, T. (1983), *Black politics in South Africa since 1945*, Longman, London.

Mamdani, M. (1996), *Citizen and subject: contemporary Africa and the legacy of late colonialism*, Princeton University Press, Princeton, N.J.

Manzo, K.A. (1996), *Creating boundaries: the politics of race and nation*, Lunne Rienner, Boulder.

Marks, S. (1986), *The ambiguities of dependence in South Africa: class, nationalism and the state in twentieth century Natal*, John Hopkins University Press, Baltimore.

Marks, S. (1995), *The tradition of non-racialism in South Africa, Eleanor Rathbone Memorial Lecture*, Liverpool.

Marks, S. and Trapido, S. (1987), 'The politics of race, class and nationalism', In S. Marks and S. Trapido (eds), *The politics of race, class and nationalism in twentieth century South Africa*, Longman, London, pp. 1–70.

Mbeki, T.M. (1998a), *Statement of Deputy President Mbeki at the opening of the debate in the National Assembly on Reconciliation and Nation Building*, 29 May, Cape Town.

Mbeki, T.M. (1998b), *Speech by Deputy President Thabo Mbeki opening the debate on the Establishment of the Commission for the Promotion and Protection of the Rights of Cultural, Religious and Linguistic Communities*, 4 August, Cape Town.

Mbeki, T.M. (2000), *Speech by Thabo Mbeki at the opening session of the National Conference on Racism*, 30 August, Johannesburg.

Meli, F. (1988), *South Africa belongs to us: a history of the ANC*, Zimbabwe Publishing House, Harare.

Munro, W.A. (1995), 'Revisiting tradition, reconstructing identity? Afrikaner nationalism and political transition in South Africa', *Politikon*, vol. 22, pp. 5-33.

O'Meara, D. (1983), *Volskapitalisme: class, capital and ideology in the development of Afrikaner nationalism*, 1934-1948, Cambridge University Press, Cambridge.

Ramutsindela, M.F. (1998), 'Afrikaner nationalism, electioneering and the politics of a volkstaat', *Politics*, vol. 18, pp. 179–188.

Ramsamy, E. (1996), 'Post-settlement South Africa and the national question: the case of the Indian minority', *Critical Sociology*, vol. 22, pp. 57-77.

Simpson, M. (1994), 'The experience of nation-building: some lessons for South Africa', *Journal of Southern African Studies*, vol. 20, pp. 463–74.

Smith, A.D. (1976), 'Introduction', In A.D. Smith (ed.), *National movements*, Macmillan, London, pp. 1–30.

Sobukwe, M.R. (1959), *Speeches of Mangaliso Sobukwe from 1949-1959 and other documents of the Pan-Africanist Congress of Azania*, PAC Observer Mission, New York.

South Africa. (1983), *Constitution of the Republic of South Africa*, Government Printer, Pretoria.

Swan, M. (1987), 'Ideology in organized Indian politics, 1894-1948', In S. Marks and S. Trapido (eds), *The politics of race, class and nationalism in twentieth century South Africa*, Longman, London, pp. 182–208.

Terreblanche, S. and Natrass, N. (1990), 'A periodization of the political economy from 1910', In N. Natrass and E. Ardington (eds) *The political economy of South Africa*, Oxford University Press, Cape Town, pp. 6–23.

Tutu, D. (1995), *The rainbow people of God*, Bantam Books, London.

Walshe, P. (1970), *The rise of African nationalism in South Africa*, Hurst & Co, London.

4   De-racializing the Land

Introduction: the land of birth

> In the South, land was not regarded as a commodity, but formed a
> fundamental part of the community's universe and sense of identity in
> material and spiritual terms (Simon, 1993, p. iv)

In the preceding chapter I have shown how national identities had been
contested over the years, and how attempts have (and still are being) made
to construct a non-racial South Africanism. There are also contests over
land. After all, land was central to the struggle for liberation in South
Africa (Benson, 1985; Letsoalo, 1992; Beinart, 1994), the more so because
the corollary to the construction of the white polity was the land
dispossession of the black people.

Conceptually, the division of land between blacks and whites
became a concrete pillar of the bifurcated state referred to in Chapter 2. In
other words, the racial division of land resonated with ideals of a divided
nationhood. Lonsdale (1992) noted that, the appropriation of land by the
colonial state set the scene for boundaries in East Africa, and Kenya in
particular. That seems to have been a general trend in colonial states.
'Throughout southern Africa', O'Laughlin (1995, p. 100) wrote, 'there is a
common historical pattern: the colonial states intervened in rural property
relations and limited access of black rural people to land as part of cheap
labour policies based on migrant labour and divided household.'

While it is common knowledge that land encompasses physical and
non-material dimensions, the analysis of the physical aspect has at most
received an upper hand in academic parlance. Such (over)emphasis is
dominant in studies on land erosion, land use patterns, and so on. The
general tenor of studies on land reform has been to measure the results of
(land reform) programmes by the acreage that has or has not been
distributed. Two clusters of reasons can be offered for that emphasis. First,
the process of land dispossession in South Africa has a history of the 'size
of the lost land'. For example, 'under the Natives Land Act of 1913 some
8.9 million hectares were defined as Native Reserves', while the 'Native
Trust and Land Act [of 1936] ... provided for the extension of black areas

39

by some 6.2 million hectares' (Christopher, 1994, pp. 32, 34). Second, the targets of land reform are often given in numerical values. The post-apartheid government 'promised' to redistribute 30 per cent of South Africa's agricultural land by 1999, for instance. All these solicit a technocratic approach to the assessment of land reform. Hence to date, most studies on land reform in South Africa have been concerned with the 'number of hectares'. Of course, land reform hinges on the amount of land that has – or has not – changed hands. However, attention should be given to other dimensions of that process. There is a current of scholarship that seeks to understand land reform and its (lack of) impact on gender relations, democracy, economy, and so on. The purpose this chapter is to show that land restitution in South Africa offers a context in which old and new interests are contested and negotiated among stakeholders. Generally, land reform is bound to generate tensions (De Wet, 1997). However, the social cost of incomplete or delayed land reform is equally high. This is evident in the formation of the Guerrilla Army of the Poor in Guatemala in 1972 (Binswanger and Deininger, 1993), and recent (i.e. 1999/2000) land invasions in Zimbabwe.

Confronting the effects of the Land Acts

The manner in which the Natives Land Act of 1913 and the Bantu Trust and Land Act of 1936 formed a cornerstone of racial land ownership patterns in South Africa are fairly well known and need no rehearsal here. Suffice it to say, those Acts were more than measures of appropriating land, but fed into the tapestry of the construction of the exclusive white polity. The Acts were, to invoke Mamdani's (1996) notion of the nature of the colonial state in Africa, concrete pillars of the 'bifurcated state.' That is, the racial division of land bolstered clear boundaries between 'citizens and subjects.' Furthermore, land ownership and settlement patterns in the former bantustans contributed to the tribalization of the pseudo states.

It follows that the process of restitution – and land reform as a whole – should be measured in terms of its impact on reconstruction at national and local scales. Nationally, the effects of the repeal of the notorious Land Acts should be evaluated broadly in terms of how far they contribute to de-racializing the land, and to changing the Janus-face of the state. Symbolically, the language of restitution as codified in the classification of urban and rural land claims reproduces that of a divided state. The post-apartheid government attempted to de-tribalize land ownership patterns through the formation of the Communal Property

Associations (CPAs). The Communal Association Bill of 1995 defined the community as a group of persons, including a tribe or potions of a tribe (South Africa, 1995). The concept of a tribe was thrown away when the Bill became the Act in 1996. On the whole, the CPAs represent a departure from the tradition of leaving the administration and/or ownership of land in the hands of chiefs. Its de-tribalizing principle is underwritten by the recognition of the community – not a 'tribe' – as a group of persons who wish to have rights to or in particular territory (South Africa 1996). According to Letsoalo (Interview, 22 May 1998, Pietersburg), the concept of the CPA has its roots in the search for a legal entity by NGOs to represent displaced communities prior to 1994, and developed within the context of an anti-chieftaincy lobby. The net effects of the CPAs would be to transform communal land ownership patterns in rural South Africa. Proponents of the land market are opposed to communal land ownership, arguing that, 'communal lands throughout the world are rarely productive in agricultural terms. In Africa, they reek of poverty, minimal motivation and a degraded natural environment' (South African Institute of Race Relations (SAIRR), 1995, p. 9). They also argue that there would be greater prosperity for all concerned if tribal chiefs were paid off and city dwellers could cash in on their rights to communal land (SAIRR, 1997). Philip Lloyd went to an extent of invoking the Duke of Westminster to argue that, 'if land rights in the old homelands were to become transferable, the tribal chiefs would become relatively wealthy overnight because they would be given a slice of cake in return for giving up their right to allocate land' (SAIRR 1997, p. 20). In contrast to this perception, communal land tenure is not inherently inferior, the intensity of problems (e.g. erosion) that are often associated with that system are inextricably linked to the history of land dispossession and allocation (Simon, 1993).

Perceptions about land use often influence land reform policies. Though South Africa's land reform policies were initially conceived in terms of redress, a significant shift from that conception became apparent after the June 1999 election. The Mbeki government introduced a new land reform policy in favour of creating black commercial farmers. Nevertheless, the idea of land restitution is still very strong. As will be shown below, restitution creates dilemmas and opportunities.

Harmonizing competing interests

On the eve of the twenty-first century, South Africa's transition to democracy earned its place among the 'miracles' of the twentieth century

(Marais, 1998). In a sense, many analysts see the country as providing a model for a peaceful resolution of conflict in deeply divided societies. Notably, the idea of the South African model is premised on the success of a negotiated political settlement and that of the first non-racial democratic elections of 1994. However the notion of the uniqueness of the South African situation and it being used as a model have come under sharper scrutiny (Mamdani, 1996; Guelke, 1996). Guelke (1996) has dismissed the notion that the 1994 national election was a unique model (miracle) for accommodating non-racialism. To him non-racialism is not a unique accommodative means to South Africa because such means have been used in Namibia, Zimbabwe and Kenya.

There could be quarrels about how far the transition from apartheid could be a model for countries in the South and beyond. However, much of what South Africa could offer as a model depend very much on the pathways and results of national programmes that aim to transform the state and society. One such programme is land reform that aims to significantly change racial patterns of land ownership (South Africa, 1997). Debates on the nature and results of that programme notwithstanding (see Levin and Weiner, 1996; De Wet, 1997; Williams, 1996; Williams et al. 1998), questions have been raised about how post-apartheid governments would balance the intended goals of land reform and other national priorities and interests. Seeking such a balance has been 'a common enough dilemma in other African countries' (Pankhurst, 1996, p. 1). This raises the question of the model(s) that would or could be adopted to harmonize national and local interests over land issues. The Makuleke land settlement of 1998 has been hailed by the media, Non-Governmental Organization (NGOs) and in government circles as a model for balancing national and local interests on restitution and environmental concerns.

Makuleke, Kruger and the diamonds

In 1997, the Chief General of the South African National Defence Force (SANDF), George Meiring, described the Makuleke and environs as an 'area riddled with claims, [many] of the claims [coming] from people who visited it once in a blue moon (*WildNet Africa*, 9 May 1997). In retrospect, those many claims are a manifestation *par excellence* of the existence of various and conflicting interests in that area. Since the 1830s when the Makuleke settled at the Pafuri triangle (Figure 4.1), the area has been subjected to different interests and clashes. During that period, the

Makuleke and the Van Rensburg-led voortrekkers clashed in the area, resulting in the annihilation of the trekkers. As Dicke (1926, pp. 1006, 1021) put it,

> [it was] in the Zoutpansberg [where] the first great sacrifice was exacted from the voortrekkers, and the first blood price for their enterprise was paid ... Chief Shinhambane Maluleke Hlekana, [is] the man who exacted the first great blood price paid by the intrepid voortrekkers.

The discovery of gold in the Witwatersrand in the 1880s also led to other interesting developments in Makuleke (also referred to here as the Pafuri). The gold mining industries created labour demand that resulted in the recruitment of labour beyond the boundaries of the Union – and later the Republic. To facilitate the recruitment of the much-needed labour, 'the Witwatersrand Native Labour Association of the Chamber of Mines established a post immediately south of the confluence of the Limpopo and the Levubu, and rapidly established monopoly over the recruitment of Africans entering Pafuri (Harries, 1987, p. 98).

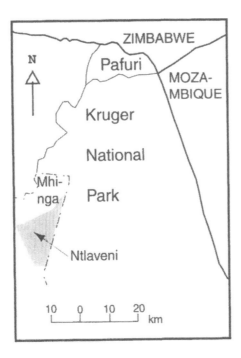

Figure 4.2   Pafuri Triangle

However, the use of the Pafuri as a labour transit was exploited by unlicensed recruiters – the crooks. As Harries (1987, p. 97) has observed, 'labour recruiters were attracted to the region by "caravans of blacks" drawn from the north of the Limpopo, who used the old trade route as a means of entering the Transvaal.' In this sense, both the licensed and unlicensed labour recruiters had a stake in the labour trade.

More crucially, it was the creation of the Kruger National Park (KNP) in March 1898 and the attempts to extend the same to the Pafuri that unleashed conflicts of interest in and over the area. On the one hand, the government aimed to extend the KNP for the protection of game. On the other hand, white farmers wanted game reserves to be opened for grazing, while land owners wanted to control hunting in their private farms (Carruthers, 1995). At the same time, the Makuleke hunters were declared poachers (Harries, 1987) (Table 4.1). Furthermore, the Department of Lands wanted the reserve for white settlement, while that of Native Affairs wanted land for the relocation of black people. For its part, the Department of Agriculture did not support the extension of the KNP into Pafuri because it feared that incorporation would lead to the spread of foot and mouth disease/rinderpest (Mouton, 1996). It is against this background that attempts by the Transvaal Provincial Authority to proclaim Makuleke as a game reserve on 26 July 1933 failed. That failure is also ascribed to the resistance to relocation by the Makuleke.

Table 4.1 Number of Poachers in Pafuri, 1958-1962

Financial Year	White Poachers	Non-White Poachers
1958/9	2	171
1959/60	2	108
1960/61	4	52
1961/2	3	49

Source: Adapted from Gilfillan, 1963, p. 186

Other interests were discernible during the creation of bantustans in the 1960s. The National Party government's system of so-called tribal authorities and its culmination in the creation of bantustans required the grouping of 'tribal' areas. The Makuleke were required to fit into the 'tribal' jigsaw puzzle. Mouton (1996, p. 6) observed that on 6 November 1957, the secretary of Native Affairs 'declared that all inhabitants of the Pafuri area ... were illegal occupants and will have to return to their homelands.' The Makuleke were considered 'Tsonga-speakers who were cut off from their ethnic "homeland" by Vendaland in the west and the Kruger National Park in the south' (Harries, 1987, p. 107). Unsurprisingly, the Makuleke were forcibly removed to Ntlaveni in 1969 – there to be put under Chief Mhinga in the district of Malamulele in the former Gazankulu. Their relocation to that area created tension and animosity between Chiefs Mhinga and Makuleke.

It is on the basis of that removal that the Makuleke were able to lodge a land claim after 1994, when the process of land restitution was in place. However, before the land claim was lodged, there were contests over the future of the Pafuri triangle. It could be suggested that those contests were crucial towards the settlement of the land claim. The discussion below will shed more light on this.

'We cannot live by diamonds alone'

As we have noted, apartheid government departments have had differences over the Pafuri, the land of the Makuleke. The extension of the KNP into Pafuri did not put those differences to rest. Instead they continued to simmer and to erupt under different circumstances. Such differences were visible over the issue of mining in Pafuri. In the 1980s, the possibilities of coking coal by Iron and Steel Corporation (Iscor) in Pafuri raised debates about the land use in the KNP. Iscor's view was and still is that steel production is of vital importance to South Africa, hence the need for coking in the KNP. It wanted to use the KNP to increase its coal reserves.

In objecting to coking, the Wildlife and Environment Society of South Africa (WESSA) (n.d., p. 4) argued that,

> if mining is permitted, Kruger National Park will be lost. Not only would the go-head mean breaking South Africa's National Parks Act, but it would mean the violation of the first principle of a national park. KNP would have to be renamed Kruger Game Reserve. It would no longer have

the international standing it has now. And if a mine is allowed, then why not grow sugar cane in the southern regions?

The anti-mining campaign in Pafuri intensified in the 1990s. The Defence Force that claimed to be a manager of the area lifted its ban on mineral prospecting in the area in 1994. It could reasonably be assumed that the army lifted the ban because of a peaceful transition to a democratic order – it was no longer important to protect the area from the 'invasion' by liberation armies. However, the unbanning of mineral prospecting gave the Department of Mineral and Energy the opportunity to issue prospecting permits to mining companies (Table 4.2). The issuing of such permits to Madimbo Mining Corporation, in particular, raised concerns among environmentalists and other government departments. They viewed the permit as a license to destroy the KNP.

The national Department of Environmental Affairs and Tourism opposed prospecting in the vicinity of the Pafuri. Its Principal Environmental Officer, Danie Smit (1996, p. 13), invoked the 'environment-for-future-generation logic' and wrote that:

> It will be a sad day indeed when we will have to face the upcoming generation with the fact that we did nothing to prevent the destruction of the last pristine wilderness areas in the southern Africa: flood-plains and pans of the Limpopo and Luvuvhu Rivers.

Both the national and provincial departments of Environment and Tourism joined forces to mount a campaign against mining. These departments proposed that the area be used for eco-tourism. The National Parks Board (NPB) – presently known as the South African National Parks – also joined the chorus on the area it saw as the subcontinent's last true wilderness about to be mined and destroyed. The Chief Executive Officer of the NPB at the time, Robbie Robinson, argued that mining will cause more destruction than the benefits it can yield (*WildNet Africa*, 18 November 1996).

Inevitably, the mining company was bound to engage with environmental concerns if it were to elicit support for mining. It promised to ensure that prospecting and mining operations will follow strict environmental guidelines. For instance, the chairperson of Madimbo Mining Corporation, Richard Bluett, maintained that,

> most of the sites where we believe the diamonds are deposited are covered by Mopani scrub which we will easily be able to restore. We won't take

out boabab forests nor will we mine where there is rivirine forest and other sensitive ecological systems (cited in *Electronic Mail & Guardian*, 18 August 1995)

Table 4.2 Mining Licences Adjacent to KNP

Claim No.	Total Claims	Claim Licence Holder
30628	25	Pafuri Metals CC
30626	50	Pafuri Metals CC
30566	50	Pafuri Metals CC
30708	46	Pafuri Metals CC
30715	25	Pafuri Metals CC
30488	10	Pafuri Metals CC
21983	46	Giant Reefs Gold Mining Co.
36672	20	G S de Vries
36673	50	G S de Vries
TOTAL	322	

All in all, mining prospecting in Pafuri brought the Department of Mineral and Energy Affairs into loggerhead with the NPB, the Department of Environmental Affairs and Tourism, the Defence Force and NGOs. That situation contributed to bringing the former Deputy Environment Minister, Peter Mokaba and Energy Affairs Minister Penuell Maduna to a review process that aimed to reevaluate the mining prospects and to give a hearing to all stakeholders. Mokaba is reported to have said that, as Deputy Minister of Environmental Affairs and Tourism, I have held discussions with Minister Maduna to request that he suspends the decision taken by his Department until a thorough strategic environmental assessment and studies to determine the best land-use options in the area have been carried out. It has been agreed that mining in the [area] cannot proceed in the light of the ecological sensitivity [there], *land claims in the area*, current proposals for developing ecotourism peace parks, and the processes and procedures by which the company obtained a mining permit (*WildNet Africa*, 29 October 1996).

Of significance to the theme of this chapter is how Minister Mokaba brought in the land claim as a factor that necessitated the reviewing of prospecting permits. As will be shown below, conservation authorities and WESSA were first skeptical about the land claim, but later employed it as a useful weapon against mining.

Fighting mining through the land claim

The Makuleke land claim (Figure 4.2) was gazetted in August 1996 when the anti-mining lobby was intensive. The Wildlife Society rejected the claim outright, arguing that,

> The granting of the land claim for the Pafuri would set a precedent for success for other land claims in conservation areas. By weakening the status of the national parks, this could set an unfortunate trend in motion of development incompatible with conservation. If our national parks, which have a high degree of protection by the law, are not sacrosanct, the outlook for conserved land in general is poor (WESSA, 1996, p. 3).

Other departments were more cautious about the land claim. The position of the Defence Force was that should the Makuleke land claim be proved valid, the army would work hand in hand with the community provided it would be allowed to carry out its security obligations (South African National Defence Force, 1996).

The NPB expressed the view that the land claim and the environmental interests in the area were competing national interests that needed proper balancing. It nonetheless advocated the use of the land for conservation. This is clear in its submission to the Land Claims Commission:

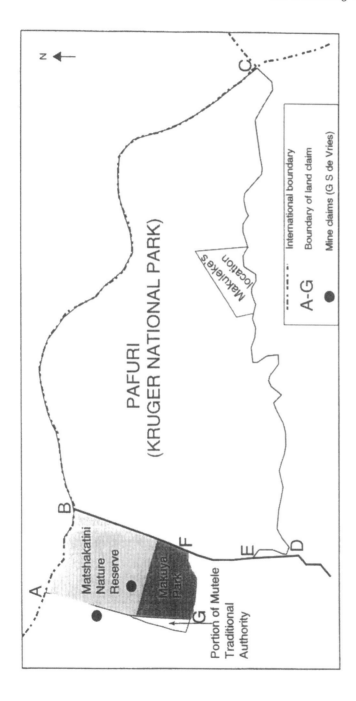

Figure 4.2 Makuleke Land Claim

> It is clear to any person involved in the settlement of this claim that two
> competing national interests are at stake. On the one hand it is in the
> national interest that communities who were deprived of their land should
> have the rights to the land restored ... On the other hand it is also in the
> national interest to conserve and expand our system of national parks ...
> The National Parks Board has for many years emphasised the
> conservation value of the Pafuri area ...' (National Parks Board, 1997, p.
> 2).

The NPB further supported the need to balance national interests by
invoking the 1996 ruling of the Land Claims Court on the Cato Manor case,
where both the restoration of land rights and development proposals were
equally compelling national interests. Whether the Cato Manor case was a
model for the Makuleke land claim settlement is beyond the scope of this
chapter. The point, though, is that conservation authorities sought to
preserve the conservation status of Pafuri, hence they intensified their
campaign against mining in that area in 1996 as I have shown. They could
obviously not use the same campaign against the land claim, because to do
so would have been sailing against the national currents. Unsurprisingly,
land claim researchers for the Wildlife Society and the Department of Land
Affairs thought to dismiss the land claim on technical grounds. The
Wildlife Society reasoned that,

> A two-thirds majority of parliament is necessary for a change in status of
> the land to be allowed, as it is a state asset. The control, management and
> maintenance of the land are vested in the National Parks Board in terms of
> the National Parks Act. The land may not be excised from the Kruger
> National Park without a resolution of parliament (Nel, 1996, p. 7).

The researcher for the Department of Land Affairs, Mouton (1996),
endorsed the Wildlife Society's view that the Makuleke be given some
form of alternative relief. However, the claimants were adamant that they
wanted back their land.

It should be noted that the land claim included the restoration of
mineral rights that were lost when the community was removed from
Pafuri. The mineral rights became one of the central issues in the land
claim. The DME had 'no objection against the claim as far as the surface of
the land is concerned, provided that the existing prospecting and mining
rights in the area are respected and protected' (Department of Mineral and

Energy, 1997, p. 1). It was of the opinion that the mineral rights must be excluded from the land claim at that stage to prevent the transfer of mineral rights to the Makuleke being in conflict with changes underway in the minerals policy. Against this backdrop, the land claim brought the issue of mining back into the agenda. The NPB and the Wildlife Society were concerned that should the Makuleke land claim succeed, the community could trade mining rights to companies they considered the enemy of conservation. There were also concerns that if mineral rights remain in the hands of the state – as Pafuri was state land at the time – the state could allow mining in future. Anti-mining campaigners such as the Wildlife Society and the NPB exploited the opportunities presented by the land claim to try and close the chapter on mining in the KNP. For instance, they proposed that the DME should withdraw prospecting permits before the land claim could be settled. This necessitated revisiting Section 7(1) of the Minerals Act of 1991 that deals with the prohibition against prospecting. Subsection 7(1)(d) of the Act reads that,

> no person shall prospect in or on the land which ... may be defined and also determined by the Minister by notice in the Gazette, except with the written consent of the Minister in accordance with such conditions as may be determined by him. (South Africa, 1991).

This section notwithstanding, anti-mining campaigners were concerned that the Minerals Act still gave the Minister the possibilities of granting consent to prospecting. They suggested that, should the mineral rights remain in the hands of the state, the state should undertake not to allow mining or prospecting in or under that land. The Makuleke, too, were to make the same undertaking should they be given those mineral rights. Clearly, the issue here was whoever owns that land should not allow mining. This condition was to be agreed to by all stakeholders before the land claim could be settled.

One can argue that the settlement of the land claim hinged on resolving the issue of mineral rights. In fact, it was one of the last issues to be resolved before the settlement was reached, and had the potential to derail the land claim. It is therefore not surprising that sufficient consensus on mineral rights was reached in April 1998, a month before a written settlement of the agreement on the land claim was entered into. The agreement was signed on 30 May 1998 and was ratified by the Land Claims Court as a Deed of Donation on 15 December the same year. According to the agreement, the Makuleke community have regained ownership of Pafuri, but had to preserve the conservation status of that area. That is, they can only use it for eco-tourism. Under the terms of the agreement, the mineral rights vest in the state. *The Sunday Times* (31 May 1998, p. 4) described the agreement as a 'historic land agreement.' De Villiers (1999) contends that the settlement may prove to be a guiding light for similar disputes in other areas of South and Southern Africa. Other sources described it as follows:

> The Makuleke/Kruger agreement sets a precedent for other land claims involving conservation areas. It also sets new parameters for situations where national parks or game reserves would like to expand by taking in underutilised ground on their boundaries (*Financial Times*, 27 March 1998, p. 52).

> 'A first for South Africa and a breakthrough for conservation' (South African National Parks, 1998, p. 1).

> 'The Makuleke community made history yesterday when the Land Claims Court in Randburg ordered that a portion of the Kruger National Parks be returned to them' (*Citizen*, 16 December 1998, p. 1).

> 'This sets a excellent precedent for land claims in other important conservation areas such as the Blyde River Canyon' (Cooper, 1998, p. 7).

The sentiment being expressed here is that the Makuleke land deal serves as a model for land claims elsewhere in the country. There is no doubt that the settlement of that land claim is a compromise from both sides. However, our understanding of what is presented as a model would be enhanced by looking at the many and varied factors that go with it.

Conclusion

In his editorial to the *African Wildlife* magazine, Cooper (1998) described the Makuleke land settlement as a historic agreement in which the community and conservation are both winners. At face value, the description is fitting because in the past the Makuleke were a looser – they lost their land rights, their stock and houses and the like – while wildlife came first. However, whether the concerns of conservationists were restoration of land rights or anti-mining is debatable. There is evidence to suggest that mining in Pafuri, rather than restoration of land rights *per se*, was a major concern of the Wildlife Society. As Cooper (1998, p. 7) wrote:

> A key concern of WESSA, raised when the *land claim* was first gazetted, was that *the coal and other minerals present in the area should not be mined*. This has been settled in the agreement as far as possible within the constraints of the current Minerals Policy governing mining activities.

This is also clear from Brown's (1998, p. 9) view that,

> WESSA was concerned that the land should be managed for conservation purposes in perpetuity, and that the use of land should take account of its status as a national asset both for tourism and conservation. We also identified the *threat that activities such as mining* and agriculture would pose to the conservation value of the land.

Obviously, what was seen to be at stake by conservationists was the conservation status of the area. That status was first threatened by mineral prospecting companies and later by the land claim. The land claim was seen to be taking away an important part of the *KNP* (*Financial Times*, 27 March 1998, p. 52). On the contrary, the land claim became useful in pushing the anti-mining campaign to its tentative conclusion – thus, preserving the status of conservation in the area. It is rather surprising that those groups that opposed the land claim came to see the settlement of that claim as a 'model'.

The Makuleke claim represents just a window on the impact and results of the broader transformation process taking place in post-apartheid South Africa. It shows attempts at harmonizing contested local and national goals that are related to the implementation of land reform. The contested interests reflected in this chapter manifest themselves in other cases - albeit in a different context - as illustrated by disputes over local government boundaries (Chapter 5).

References

Beinart, W. (1994), *Twentieth-century South Africa*, Opus, London.

Benson, M. (1985), *South Africa: struggle for a birthright*, International Defence and Aid Fund for Southern Africa, London.

Binswanger, H.P. and Deininger, K. (1993), 'South African land policy: the legacy of history and current options', *World Development*, vol. 21, pp. 1451–75.

Brown, L. (1998), 'WESSA's concerns over Makuleke claim addressed', *African Wildlife*, vol. 52, p. 9.

Carruthers, J. (1995), *The Kruger National Park: A social and political history*. University of Natal Press, Pietermaritzburg.

Citizen. (1998), 16 December, p. 1, Johannesburg.

Cooper, K. (1998), 'Editorial', *African Wildlife*, vol. 52, p. 7.

Christopher, A.J. (1994), 'Indigenous land claims in the Anglophone world', *Land Use Policy*, vol. 11, pp. 31–44.

De Villiers, B. (1999), *Land claims and national parks: the Makuleke experience*. Pretoria: Human Sciences Research Council.

De Wet, C. (1997), 'Land reform in South Africa: A vehicle for justice and reconciliation, or a source of further inequality and conflict?' *Development Southern Africa*, vol. 14, pp. 355–62.

Dicke, B.H. (1926) 'The first voortrekkers to the Northern Transvaal and the massacre of the Van Rensburg Trek', *South African Journal of Science*, vol. 23, pp. 1006-1021.

Department of Minerals and Energy. (1997), *Submission to the Department of Land Affairs*, Pretoria.

Electronic Mail & Guardian. (1995), 18 August, Johannesburg (www.mg.co.za).

Financial Times. (1998), 27 March, p. 52, Johannesburg.

Guelke, A. (1996) 'Dissecting the South African miracle: African parallels', *Nationalism and Ethnic Politics*, vol. 2, pp. 141–54.

Harries, P. (1987) 'A forgotten corner of the Transvaal': reconstructing the history of a relocated community through oral testimony and song' In B. Bozzoli (ed), *Class, community and conflict: southern African perspectives*, Ravan Press, Johannesburg, pp. 93–134.

Letsoalo, E.M. (1992), *Restoration of land rights: problems and prospects*, Paper presented to the Southern African Political Economy Series (SAPES) Conference, 18-22 August, Cape Town.

Levin, R. and Weiner, D. (1996), 'The politics of land reform in South Africa after apartheid: perspectives, problems and prospects', *Journal of Peasant Studies*, vol. 23, pp. 93–119.

Lonsdale, J. (1992), 'The conquest state of Kenya, 1895-1905', In B. Berman and J. Lonsdale, *Unhappy valley (Book Two)*, James Currey, London, pp. 13–44.

Mamdani, M. (1996), *Citizen and Subject: contemporary Africa and the legacy of late Colonialism*, Princeton University Press, Princeton, N.J.

Marais, H. (1998), *South Africa Limits to Change: the political economy of transformation*, Zed Books, London.

Mouton, B.F. (1996), *Makuleke's location, district of Zoutpansberg, Northern Province. Report No 115/95*, Department of Land Affairs, Pretoria.

National Parks Board. (1997), *Submission to the Regional Land Claims Commissioner*, Pretoria.

Nel, M. (1996), 'Kruger Land Claim: South Africa's premier national park faces a major challenge', *African Wildlife*, vol. 50, pp. 6–9.

O'Laughlin, B. (1995), 'Past and present options: land reform in Mozambique', *Review of African Political Economy*, vol. 63, pp. 99–106.

Pankhurst, D. (1996), *A resoluble conflict? The politics of land in Namibia*. Peace *Research Report No 36*, University of Bradford.

Simon, D. (1993), 'The communal land question revisited', *Third World Planning Review*, vol. 15, pp. iii–vii.

South Africa. (1991), *Minerals Act*, Government Printer, Pretoria.

South Africa. (1995), *Communal Association Bill*, Government Printer, Pretoria.

South African Institute of Race Relations. (1995), 'Freehold key to land reform', *Frontiers of Freedom*, vol. 5, pp. 7–9.

South African Institute of Race Relations. (1997), 'Give the people their land', *Frontiers of Freedom*, vol. 12, pp. 18–20.

Smit, D.W.J. (1996) 'Leave this "diamond" unmined', *Conserva*, vol. 12, pp. 11–13.

South African National Defence Force. (1996), *Submission to the Department of Land Affairs*, Pretoria.

South African National Parks. (1998), Media Release, 25 March, Pretoria.

Sunday Times. (1998), 31 May, p. 4, Johannesburg.

Wildlife and Environment Society of South Africa. (n.d), 'The fight against mining in the Kruger National Park', *African Wildlife (Special Issue)*.

Wildlife and Environment Society of South Africa. (1996), *Submission to the Regional Land Claims Commissioner*, Pretoria.

WildNet Africa. (1996), 29 October, wildnetafrica.oc.za.

WildNet Africa, (1996), 18 November, wildnetafrica.co.za.

WildNet Africa. (1997), 9 May, wildnetafrica.co.za.

Williams, G. (1996), 'Setting the agenda: A critique of the World Bank's rural restructuring programme for South Africa', *Journal of Southern African Studies*, vol. 22, pp. 139–66.

Williams, G., Ewert, J., Hamann, J. and Vink, N. (1998), 'Liberalizing markets and reforming land in South Africa', *Journal of Contemporary African Studies*, Vol. 16, pp. 65–94.

5 Reconstructing Urban Spaces: A Disputed Terrain

Introduction: towards urban reconstruction

In this chapter attention is given to local government restructuring in order to highlight the impact of transition on locality. In the conceptual terms set out in Chapter 2, this chapter looks at the process of transforming local governance in line with the vision of a non-racial society. As I have argued in that Chapter, one of the tasks of the post-colonial state was to address the problems of internal divisions that had been colonially institutionalized through local apparatus.

In the case of South Africa, local government played a significant role in the administration of apartheid. No wonder that that institution was highly centralized and managed within a system of institutional patronage. More crucially, segregated local government represented the terrain on which the unequal distribution of resources was launched and sustained by the apartheid state. To that end, the transformation of local government became a political imperative in the post-apartheid era. Constitutional changes in the 1990s, and the Local Government Transition Act of 1993 in particular, set the stage for local government restructuring; the aim being to make the institution compatible with the vision of a non-racial order.

The transformation of local government is highly contentious in urban areas, not least because towns and cities in South Africa were a bastion of white power. In this context, challenges arising from the transformation of the urban system are a reflection of the total complex of those in society as a whole (Harvey, 1996). Nevertheless, cities may not necessarily be reflectors of larger societal tensions and dynamics because they can also exert their own independent effect on society. Nonetheless, South African cities provide the worse-case scenario of racial segregation and socio-economic inequality (Parnell, 1997). There is no space here to rehearse the case of urban planning under apartheid. It is important, though, to note that planning was couched in terms of anti-urbanism and assumed a technocratic stance (Turok, 1994; Dewar, 1995; Mabin, 1995).

Anti-urbanism is of course not a unique South African phenomenon; it prevailed in the West as a counterweight to the perceived or real chaos brought to cities by the Industrial Revolution (Keith and Cross, 1993). However, in South Africa it assumed a peculiar character under a plethora of racist laws. There anti-urbanism was directed specifically to Africans, coloureds and Indians who were seen through racist lenses as a social pest. As Bickford-Smith (1989, p. 49) commented:

> Throughout the nineteenth century, social mobility of other than whites threatened the dominant belief that social order should coincide with racial order, if not be actually achieved via the latter.

The new political dispensation aims to dismantle that racial order by changing the apartheid city through spatial integration, the restructuring of urban governance and the redistribution of resources. However, strategies to achieve those goals have been a bone of contention since 1993, as illustrated by case studies in Johannesburg and Groblersdal.

Johannesburg: sharing the city of gold

Attempts to transform the city of Johannesburg can conveniently be traced to the creation of Regional Services Councils (RSCs) in the early 1980s. The RSCs were conceived by the government as one of the strategies to 'normalize' local authorities and 'a means of institutionalizing discussion and dialogue at the local level' (Humphries, 1991, p. 80). The RSCs sought to bring together representatives of white, African, coloured and Indian local government structures and to upgrade socio-economic conditions in the townships through resource redistribution (Humphries, 1991). However, the RSCs were also an attempt to deal with a political problem as shown by the following statement:

> The present constitutional dispensation gives some communities insufficient say in joint decision-making at the third tier. We shall have to bring this [RSCs] to be able to offer effective resistance to internal and external groups who wish to overthrow the dispensation in South Africa violently. We shall have to do this in an attempt to accommodate the spiral rising expectations, especially in black townships (cited in Seethal, 1991, p. 14).

However, the demarcation of RSCs for the Witwatersrand in 1986 served to open the debate on the creation of a metropolitan area for Johannesburg. The Demarcation Board for the RSCs viewed the Witwatersrand as 'a fairly well integrated and economically interdependent area, its fabric [being] well structured and almost continuous in its urban development' (Mabin, 1999, p. 163). Though the Johannesburg City Council supported the idea of a single RSC for the Witwatersrand, the provincial government favoured the separation of townships from central Johannesburg. This is not surprising because in that period (i.e. the 1980s) the apartheid government was only interested in making cosmetic changes. Furthermore, the need to implement apartheid at all levels of the state meant that local authorities were prevented by central government from implementing structural changes that were at variance with the overall apartheid ideology. For the Witwatersrand this meant that decisions by the Demarcation Board and the Johannesburg City Council to pursue integration in the RSCs was against the apartheid policy of separate development. However, political changes in the 1990s opened space for profound local government restructuring.

In 1991 the National Party (NP) promulgated the Interim Measures for Local Government Act (hereafter IMA). The main objective of this Act was to provide the framework for local negotiations. As Turok (1994, p. 356) has noted, 'the government passed this Act to encourage neighbouring local councils to cooperate in providing services where they had broken down, or to undertake more substantial reorganisation, including incorporation or amalgamation, where local councils had lost credibility.' Though the IMA created a forum for local negotiations, it,

> included elements clearly designed to protect white or middle-class minorities: ward councils to preserve the character of local communities and provide them with a say regarding their residential environment, and a dual voting role which would have given half the voting strength to property owners (Lemon, 1996, p. 37).

On the whole the IMA required local negotiations to be conducted on a voluntary basis. The results of the IMA were that only 42 of the 843 segregated local authorities had reached consensus in 1993 (Turok, 1994). The ANC rejected the IMA because it saw the Act as a desperate attempt to entrench white privileges at the local level (Maharaj, 1997). After all, it was doubtful whether the apartheid state could transform itself. In fact, the IMA permitted more conservative authorities to resist fundamental change.

In the Witwatersrand, local negotiation among representatives from local authorities and civic organizations gave the city a head start (Lemon, 1996). At that time, the city council sought to use negotiations as a step towards ending the rent boycott in the townships. The immediate result of the negotiations was the signing of the Soweto Accord on 24 September 1990. The Accord paved way for the formation of the Central Witwatersrand Metropolitan Chambers (CWMC). As the chair of the CWMC at that time, Van Zyl Slabbert, commented,

> the basic reason for the Chamber's existence [was] to provide a forum for negotiating non-racial and democratic structures of local government and to improve the quality of life of the people by establishing a common tax base and upgrading the quality of essential services' (cited in Mabin, 1999, p. 166).

Turok (1994, p. 360 has summarized the aims of the CWMC as to,

> transform the central [Johannesburg] metropolitan region so that the standard and quality of life of people can be improved, and political participation ensured by establishing a legitimate system of local and metropolitan government that is based on non-racialism, democracy and a common fiscal base, and is capable of promoting constitutional, economic, institutional, physical, financial and social development.

Thus, the CWMC sought to transform the urban geography of the city and its surrounding townships by integrating what had been fragmented (Mabin, 1995). In practice, that integration required the demarcation of boundaries for local authorities to ensure that informal settlements on the outskirts of towns and cities, and urban settlements displaced behind former bantustan boundaries are incorporated into the jurisdiction of new local authorities (ANC, 1994).

The demarcation of the envisaged region was contested because boundaries are not mere physical lines, but are embedded with, and reflect socio-political meanings and have profound impact on the distribution of power and resources. In this context, the demarcation of the Witwatersrand metropolitan area could not have been a smooth process. The debate centred on the areas to be included or excluded from the new local structure. The results of the negotiations were the adoption of the area of the Greater Johannesburg as the same to that of the Central Witwatersrand Regional Services Council, the acceptance of a strong metropolitan government and the abolition of existing local government (Mabin, 1999).

Nonetheless, the precise boundaries of the Metropolitan Substructure (MSS) remained contentious.

To this end, the Local Government Transition Act (LGTA) of 1993 gave impetus to the vision of the CWMC. The LGTA required stakeholders to form negotiating forums. Since the CWMC was already in operation, that body was recognized for the purposes of the LGTA. Thus, the Gauteng Demarcation Board finalized the boundaries of the Johannesburg Metropolitan Council in 1995, in preparation for the local government elections in November that year.

Generally, the interim local government structures were operational till 1999, when the final constitution came into effect. The final phase of local government restructuring envisages a complete transformation of the local government system in line with the final constitution of 1996 that came into effect in 1999. Chapter 7 of the final constitution considers local government as a sphere of government in its own right without interference by national and provincial governments (South Africa, 1996). It also links local government to the constitutional principle of co-operative government between tiers (Figure 5.1), the aim being to shift the roles of local government from mere administration to those of development – a developmental local government (DLG) model. For the DLG to be effective, a re-demarcation of municipalities was considered necessary. Against this background, the Municipal Demarcation Act of 1998 (South Africa, 1998a) seeks to define municipal boundaries in order to facilitate co-ordination between municipal, provincial and national functions, services and programmes; and to integrate social and economic planning and development within an inclusive tax base.

The re-demarcation of DLG implies the redefinition of municipalities. Effectively, the aim is to reduce 843 municipalities to about 285 (*Star*, 23 February 2000, p. 3) and to establish different types of municipalities. The national Demarcation Board under the chairperson of Michael Sutcliffe recommended the establishment of megacities in Johannesburg, Cape Town, Durban, Port Elizabeth, Pretoria and the East Rand. A megacity represents an enlarged single city structure that encloses different municipalities into an administrative structure with a single metropolitan economy.

The White Paper on Local Government proposes the establishment of metropolitan government on three main grounds. The first of these is that 'metropolitan government creates a basis for equitable and socially just metropolitan governance' (South Africa, 1998a, p. 59).

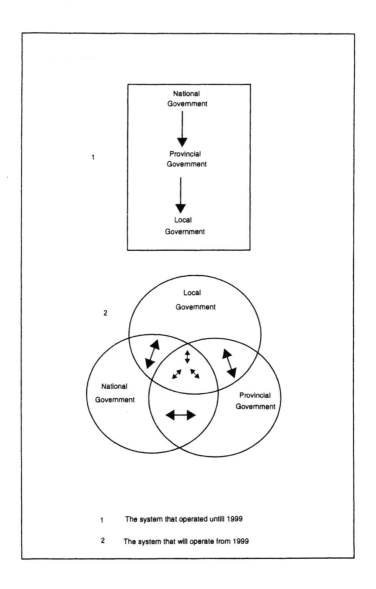

Figure 5.1 Co-operative Government

That is, the absence of metropolitan government leads to exclusionary practices as a result of metropolitan areas seeking to exclude the poor and to protect their tax base. Second, metropolitan government are seen to be a mechanism for promoting 'strategic land-use planning, and co-ordinated public investment in physical and social infrastructure. Third, such government is seen to be able 'to develop a city-wide framework for economic and social development, and enhance the economic competitiveness and well-being of the city' (South Africa, 1998b, p. 60).

While the megacity model stems from a search for an appropriate local government model (*Sowetan*, 19 September 1997), the precise nature and operation of such model has created a much heated debate. In Johannesburg two camps can be identified in that debate: one that seeks to keep separate municipalities as an effective and cheaper system of governance. The New National Party and the Democratic Party support that camp (*Electronic Mail & Guardian*, 22 September 1997). Opponents of the megacity see the model as leading to the centralization of political power; a bloated, wasteful and inefficient bureaucracy; a drop in the quality of services and property values, and so on. Generally, opponents of the model see it as thrashing the white suburbs to bring equality to the new South Africa (*Sunday Times*, 28 September 1997, p. 2).

The second camp consists of enthusiasts of a megacity and has the backing of the African National Congress (ANC). They see the megacity as model a viable option for eliminating duplication of administrative structures, the integration of the provision of services, the creation of a single vibrant metropolitan economy, and so forth. On the whole, the megacity model is seen as an effective tool in the redistribution of resources. The question of redistribution of resources is a complex one, not least because of the problems of rental payments, the privatization of service provision (Igoli 2002 proposal) and huge backlogs in the townships. The transformation of local government has also been contested in small towns such as Groblersdal.

Defending the last white frontier: the case of Groblersdal

[The] town council of Groblersdal is the only local government body in the country which was established and constituted under the previous dispensation, which is to the dissatisfaction of the majority of the inhabitants of the area who were at the time disenfranchised (Northern Province Archive, p. 2).

This statement captures the past and contemporary situation in the town and district of Groblersdal. The town of Groblersdal existed as an exclusive 'white' area in the former Transvaal Province. As a typical apartheid town, Groblersdal was physically separated from the black township of Motetema (Figure 5.2). This separation reflected the wider social engineering by which racial magisterial districts were created in order to facilitate the process of 'homelandization.'

It has already been pointed out that local government restructuring was guided by the provisions of the LGTA of 1993. The Town Council invoked Part IV of the Act to establish a forum as a pre-interim phase of local government restructuring. A meeting to establish the forum and to determine the forum area was held on 18 April 1994, after which the Groblersdal Negotiating Forum sought approval from the Transvaal

Provincial Administration (Town Council of Groblersdal, 9 May 1994). However, the forum did not get official recognition because it had excluded Africans living in areas adjacent to Groblersdal. Africans protested their exclusion. For instance, the local branch of the ANC resolved that,

> it will do all within its power to oppose recognition of the forum of 18 April1994; it will, despite the reluctance of the Forum of 18 [April] 1994, negotiate with the 'Forum' firstly to establish itself as a full member with a voting right as envisaged by the Act (ANC, 1994b).

Owing to the growing pressure from local protests by African communities and the stigma of Groblersdal as a *volkstaat*, whites in Groblersdal were persuaded to rethink the forum area. There was a concern among whites to get Africans into the forum area in line with the demand of non-racial local government in the country. The town council made attempts (but failed) to build low cost housing for 'people of all races' within its area of jurisdiction. It had a plan for 1 500 stands as part of the RDP, and also obtained R33 million from undisclosed outside sources (Confidential Interview, 10 June 1998, Pietersburg). It could be suggested that the 1 500 stands were to accommodate blacks who would possibly not pose a serious counterweight to 3 800 whites in the area in the envisaged local government elections (Table 5.1).

Other attempts by the Town Council included the expansion of the forum area to include the African township of Motetema. This did not solve the problem either, because residents in Tafelkop and Moutse also wanted to be included in the new TLC with Groblersdal.

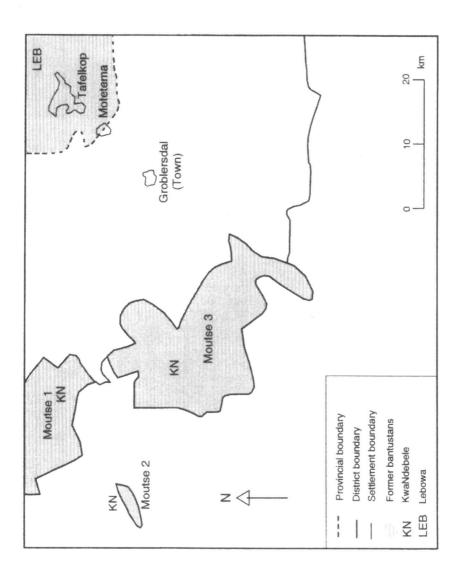

Figure 5.2 Grobersdal and Surrounding

Table 5.1 Population in and around Groblersdal, 1996

Area	Urban	Semi-Urban	Rural	Per cent
Aquaville	-	118	-	0.07
Elandskraal	20 173	-		12.14
Groblersdal	3 800	-	-	2.29
Hinlopen	-	-	1 152	0.69
Leeufontein	6 881	-	-	4.14
Marble Hall	3 600	-	-	2.17
Motetema	6 285	-	-	3.78
Regae	190	10 724		6.57
Tafelkop		55 000	1 988	34.29
Weltevreden	-	965		0.58
Zaaiplaats	8 600	7 400	16 000	19.26
Other			23 290	14.02
TOTAL	**49 529**	**74 207**	**42 430**	**100.00**

Source: Winterbach Potgieter and Partners, 1997, p.5

It should be noted that neither the provincial nor the national government could impose changes at the local level; to do so would have been unconstitutional. These tiers of government could only address the problem in Groblersdal through a combination of measures within the law. There was an attempt at the provincial level to resolve the problem through an Inter-Provincial Technical Team in early 1995 (Ramutsindela, 1998).

The Team investigated the problem and recommended the establishment of two TLCs, the first comprising Groblersdal, Motetema, Tafelkop and Moutse 2 and 3 (see figure 5.2). The second included Marble Hall, Leeufontein, Middle Lepelle and Moutse 1. The Town Council of Groblersdal rejected those recommendations, preferring a single TLC with Motetema. In the view of the Town Council, Groblersdal and Motetema were the only urban areas in the vicinity. Since the LGTA had divided TLCs into urban and rural, the Town Council considered the amalgamation of Groblersdal and adjacent rural areas as a violation of the constitution. Indeed, the legal representative (Northern Province Archive, 16 November 1994, p. 2) of the existing all-white council warned that,

> Should a council [i.e. TLC] be forced on our client [i.e. the existing town council of Groblersdal] … and it being against the wishes of our client, the Supreme Court of South Africa shall be approached for an interdict preventing the establishment of such a transitional local council.

The problem not only hinged on the urban-rural dichotomy, but was also compounded by provincial boundaries. Groblersdal was located in Mpumalanga while contesting African settlements were in Northern Province. The LTGA did not make provision for cross-border TLCs. Furthermore, the provincial boundaries had already been recognized in the Constitution.

The government determined that it would not amend provincial boundaries, all the more because there were already approximately fourteen disputes over provincial boundaries in the whole country. It should be noted that whites in Groblersdal not only preferred integration with Motetema, but also determined to be administered in Mpumalanga. White interest in Mpumalanga was both political and economic.

Before the 1994 national election, Mpumalanga had the third largest concentration of potential white voters (20.41 per cent), after Gauteng and the Western Cape which had 34.02 and 28.14 per cent respectively (Development Bank of Southern Africa, 1992). In electoral terms, Mpumalanga offered a greater potential for political parties such as the NP and Freedom Front (FF) to secure votes from the white electorate than Northern Province where the white electorate accounted for only 4.23 per cent. Such a political calculation seems to have served a purpose because the NP and FF performed better in Mpumalanga, where they won three and two legislative seats respectively, than in Northern Province where they secured one seat each. Unsurprisingly, these parties have directly (and indirectly) opposed any suggestion to relocate Groblersdal to Northern Province. It can reasonably be suggested that the majority of whites in Groblersdal identified with Mpumalanga. Most of the cars, tractors and trucks that were being driven by whites had the registration number 'MP' for Mpumalanga or 'T' for the old Transvaal (Observations, 25 September 1997, Groblersdal). Of significance are the economic activities such as game farming, commercial agriculture and so on, that are dominated by whites in that province. There are speculations that individuals who have political and economic interests in Groblersdal would like to see that area remaining in Mpumalanga (Anonymous interview, 10 June 1998, Pietersburg).

Section 7 of the LGTA provides some guidelines for dealing with areas such as Groblersdal, where a forum has not been established:

> If a forum for any reason whatsoever [has] not been established or recognized as contemplated in Section 6 on 30 November 1994, the MEC may, notwithstanding anything to the contrary contained in this Act, forthwith determine that the option referred to in paragraph (b) or (c) of Section 7 (1) shall be applied to the local government bodies in any such area and shall, within a period of *30 days* [my emphasis] of such determination, exercise the powers conferred upon him or her by Section 10(1), incorporating the provisions of such determination in the proclamation contemplated in the said section (South Africa 1993, p. 13).

The implication of this section is that the provincial Minister (MEC) should have reopened discussions on the establishment of the negotiating forum after 30 November 1994, and should have exercised his/ her powers to establish a body for the pre-interim phase in January 1995. This opportunity was lost partly because of the disputes around the forum area. The LGTA (subsection (1B)(c) of Section 9) further states that,

> any transitional council or transitional metropolitan substructure or other body of the pre-interim phase for which no election has been held before or on 31 August 1996 may be dissolved by the Minister by notice in the Provincial Gazette' (South Africa 1993, p. 14).

The Groblersdal council did not fit into any of these structures because it was neither transitional nor of a pre-interim phase, since it was not established by the Transition Act. The MEC had two options to use in order to dissolve the town council, the first being to establish a 'stand alone' TLC for Groblersdal. Undoubtedly, the town council would have supported such a step because of its unwillingness to form a TLC with surrounding African areas. A 'stand alone' TLC was problematic in that it would have been a perpetuation of (territorial and economic) racial segregation – the very problem that local government restructuring aimed to address.

The MEC's intention to establish such a 'stand alone' TLC was, of course, a contingency plan to establish a structure that could be dissolved in terms of the LGTA. There was no guarantee that the MEC would have been successful in dismantling a 'stand alone' that s/he would have created.

The second option was to proclaim a TLC consisting of Groblersdal and Motetema, and to rekindle the joint administration of the envisaged TLC by the two provinces on the basis of the inter-provincial agreement (ITT, 1995). This type of inter-provincial administration required the approval of the President in terms of Section 10A of the Local Government Transition Act of 1993. This approval was given on 22 August 1996.

Having cleared the formalities for the inter-provincial administration of Groblersdal, Northern Province could become involved in the affairs of the town. The first step was to proclaim the town of Groblersdal as part of its administrative Bushveld District through Proclamation 30 of 1996 (Northern Province, 1996). The plan was to bring the administration of Groblersdal under Northern Province so that the disputed areas of Groblersdal town, Motetema and Tafelkop should fall under one province, in order for that province to handle the disputed areas together.

The second step was to dissolve the town council and appoint new administrators on the grounds that elections were not held in that area on 31 August 1996. The Northern Province terminated the terms of office of the town council and appointed Messrs Mathabatha and Fick as administrators from 1 November 1996 (Northern Province 1996c). The new administrators were expected, *inter alia*, to start a negotiation process within the Groblersdal area with the aim of establishing a TLC.

The presence of Northern Province administrators in Groblersdal fomented protests by Africans and whites who did not want to belong to Northern Province. The pro-Mpumalanga group disrupted a rally (at Tafelkop) which was supposed to be addressed by the Northern Province MEC for Local Government, Mr R.J. Dombo, on 9 November 1996. In the same month, the new administrators were forced out of office by protesters, leading to the outbreak of violence in the area.

Protests by the town council were more of a legal nature. As the council had resolved that Groblersdal should remain in Mpumalanga, it successfully requested the Supreme Court to nullify the appointment of the new administrators (Case no. 24293/96) on 22 November 1996. One of the mistakes made by Northern Province in attempting to incorporate Groblersdal into its territory was to proclaim the town as part of that province without involving the demarcation board and the town council, as required by the Transition Act. The verdict of the Supreme Court against Northern Province was another blow to the restructuring of the local council in Groblersdal. As a result of this verdict, the town council was reinstated and the two administrators appointed by the Northern Province were withdrawn from 1 March 1997.

Subsequent efforts by Northern Province to demarcate TLCs in Groblersdal and environs were fruitless, and only served to deepen the crisis of local government restructuring in that area. For instance, the hearing by the Demarcation Board at Groblersdal was disrupted by pro-Mpumalanga groups on 15 April 1997. Residents in villages near Groblersdal rioted and gave President Mandela 14 days to respond to their demands to be incorporated into Mpumalanga (SABC, 22 May 1997). It is against the background of disputes in Groblersdal and environs that local government restructuring in the area was stalemated. This deadlock remained unresolved till 1999, when the redemarcation of developmental local government got under way. At the time of writing (i.e. October 2000) the problem of local government restructuring appeared to have been solved through a cross-border municipality that straddle the boundary between Mpumalanga and Northern Province. That new genre of municipalities has been made possible through the promulgation of the Municipal Demarcation Act of 1998.

Conclusion

The post-apartheid efforts to restructure the state in South Africa are not without precedent on the continent, because similar efforts were made by other states after bloody wars in Angola and Mozambique, or negotiated deals in Zimbabwe and Namibia (see Sidaway and Simon 1993; Simon 1996). Such efforts, at best, represent the quest by the new states to infuse new meanings to the territorial organization in line with the policies and visions of those states. Conceptually, the model of local government restructuring adopted by the post-apartheid government attempts to devolve powers from central government while, at the same time, seeking to drive local agenda from the centre. In this context, there is a perception by national government that local government on its own cannot fulfil its constitutional obligation. Correctly so, major obstacles faced by transitional local authorities not only reflected the existence of financial constraints, but illustrated lack of capacity and organizational ability to run local affairs. Under these conditions, local autonomy becomes a dubious concept that creates the inevitable interventionist strategies that could hamper the development of such autonomy. As alluded to above, conditions of this nature account for the persistence of a pseudo devolution of power to local government, especially on the African continent.

In the case of South Africa, territorial restructuring aimed to break racial and ethnic 'spatial containers' that characterized the apartheid state, in order to promote non-racialism (see Chapter 3) and the sharing of hitherto racially distributed resources. While these broad aims were pursued at national level, their impact was strongly felt at local government level. It is therefore not surprising that the local terrain was highly contested, as the cases of Johannesburg and Groblersdal illustrate.

References

African National Congress. (1994a), *The Reconstruction and Development Programme*, Umanyano, Johannesburg.

African National Congress. (1994b), *Memorandum*, 25 August, Motetema.

Bickford-Smith, V. (1989) 'A special tradition of multi-racialism'? segregation in Cape Town in the late nineteenth and early twentieth centuries', In G.J. Wilmont, and M. Simons, (eds), *The angry divide: social and economic history of the Western Cape*, David Philip, Cape Town, pp. 47–62.

Development Bank of Southern Africa. (1992), *Regional distribution of potential voters in South Africa*, Halfway House.

Dewar, D. (1995), 'The urban question in South Africa: the need for a planning paradigm shift', *Third World Planning Review*, vol. 17, pp. 407–418.

Electronic Mail & Guardian. (1997), 22 September, Johannesburg.

Harvey, D. (1996), 'On planning the ideology of planning', In S. Campbell and S. Fairnstein (eds), *Reading in Planning Theory*, Blackwell, Oxford, pp. 176–97.

Humphries, R. (1991), 'Wither Regional Services Council', In M. Swirling, R. Humphries and S. Khehla (eds), *Apartheid city in transition*, Oxford University Press, Cape Town, pp. 78–90.

Interprovincial Technical Team. (1995), *Report of the discussions held by members of the Interprovincial Technical Team*, 25 April, Groblersdal.

Keith, M. and Cross, M. (1993), 'Racism and the postmodern city', In M. Cross and M. Keith (eds), *Racism, the city and the state*, Routledge, London, pp. 1–30.

Lemon, A. (1996), 'The new political geography of the local state in South Africa', *Malaysian Journal of Tropical Geography*, vol. 27, pp. 35–45.

Mabin, A. (1995), 'On the problems and prospects of overcoming segregation and fragmentation in Southern African cities in the postmodern era', In S. Watson and K. Gibson (eds) *Postmodern cities and space*, Blackwell, Oxford, pp. 187–98.

Mabin, A. (1999), 'From hard top to soft serve: demarcation of metropolitan government in Johannesburg', In R. Cameron, *A Tale of three cities*, Van Schaik, Pretoria, pp. 159–200.

Maharaj, B. (1997), 'The politics of local government restructuring and apartheid transformation in South Africa', *Journal of Contemporary African Studies*, vol. 15, pp. 261–285.

Northern Province. (1994), *Unclassified documents*, 16 November, Pietersburg.

Northern Province. (1996), *Proclamation 30 of 1996*, Pietersburg.

Northern Province Archive, *LHL20/9/6/5/26*, Pietersburg.

Ramutsindela, M.F. (1998), 'The survival of apartheid's last town council in Groblerdal, South Africa', *Development Southern Africa*, vol. 15, pp. 1–12.

Parnell, S. (1997), 'South Africa's cities: perspectives from the ivory tower of urban studies', *Urban Studies*, vol. 34, pp. 891-906.

Seethal, C. (1991), 'Restructuring the local state in South Africa: Regional Services Councils and crisis resolution', *Political Geography Quarterly*, vol. 10, pp. 8–25.

Sidaway, J.D. and Simon, D. (1993), 'Geopolitical transition and state formation: the changing political geographies of Angola, Mozambique and Namibia', *Journal of Southern African Studies*, vol. 19, pp. 6–28.

Simon, D. (1996), 'Restructuring the local state in post-apartheid cities: Namibian experience and lessons for South Africa', *African Affairs*, vol. 95, pp. 51–84.

South Africa. (1993), *Local Government Transition Act*, Government Printer, Pretoria.

South Africa. (1998a), *Municipal Demarcation Act*, Government Printer, Pretoria.

South Africa. (1998b), *White Paper on Local Government*, Government Printer, Pretoria.

South African Broadcasting Corporation. (1997), 22 May, Auckland Park.

Sowetan. (1997), 19 September, Johannesburg.

Town Council of Groblersdal. (1994), *Application for the recognition of Groblersdal Negotiating Forum*, 9 May, Groblersdal.

Turok, I. (1994), 'Urban planning in the transition from apartheid: part 2 – towards reconstruction', *Town Planning Review*, vol. 65, pp. 355-74.

Wniterbach Potgieter and Partners. (1997), *Report on the demarcation of the area comprising the Magisterial District of Groblersdal and the areas of Middle Lepelle, Leuwfontein, MoMotetema and Tafelkop*, Pietersburg.

6 The Legacy of Traditional Authorities

Introduction: what future for traditional authorities?

Like most African states, South Africa has a legacy of the rural-urban dichotomy that had been bolstered by administrative structures. As Mamdani (1996) has noted, colonialism created a bifurcated state that entrenched the separation of rural and urban areas. That duality was anchored on the separation of urban power from rural power. Post-independence African leaders therefore were faced with the legacy of that duality. Attempts to transform the spatiality of the colony led to two categories of states: conservative and radical. Conservative states such as Lesotho retained local state apparatus such as chiefs and traditional local councils. Radical states such as Tanzania abolished the institution of traditional leaders after independence. In some cases there have been attempts to amalgamate traditional and modern institutions.

Thus, there is an impasse on approaches to Africa's predicament. Mamdani (1996, p. 3) commented on that impasse as follows:

> For modernists, the problem is that civil society is an embryonic and marginal construct in Africa; for communitarians, it is that real flesh-and-blood communities that comprise Africa are marginalized from public life as so many 'tribes.' The liberal solution is to locate politics in civil society, and the Africanist solution is to put Africa's age-old communities at the centre of African politics.

In the present-day global environment, neither of the two perspectives would successfully take the continent forward. On one hand, the route back to Africa's age-old communities could prove very hard to reconstruct. On the other hand, western systems and values have not always been good for the continent and its people. Admittedly, both perspectives have merits and disadvantages that need a rethink. As will be shown below, South Africa, too, has to find its way through that impasse. A case in point is on the issue of traditional authorities.

Apartheid traditional authorities

As in other colonial settings, the invention of traditional authorities in South Africa formed a strand in the tapestry of the politics of colonial rule. The tradition that colonial powers privileged as the customary was the one with the least historical depth, the aim being to use it in the day-to-day violence of the colonial system (Mamdani, 1996). As an ideological construct, successive white minority governments incorporated traditional authorities into the body politic. In South Africa, the Native Administration Act of 1927 conferred 'civil' jurisdiction on chiefs and attempted to establish administrative uniformity throughout the country (Hendricks and Ntsebeza, 1999). That Act was not only meant for administrative purposes, but also served the political goal of separating rural Africans from 'white' South Africa. No wonder that the government of the day was virtually obsessed with chiefs and tribes. Two years before the promulgation of the Native Administration Act (i.e. in 1925), The Native Affairs Department (NAD) established an ethnological sections, the main aim being to serve 'more practical ends of native administration' (Ritchken, 1994, p. 65). Officers in the NAD were encouraged to take vacation courses in ethnology and social anthropology, while Schapera's study of Tswana law and customs became the textbook for commissioners (Ritchken, 1994). Scholars supportive of the ideology of apartheid fed the government with the much-needed information on the nature of chiefs and tribal government.

All these helped shape the definition of the 'tribe' and in the implementation of native policies. Ritchken (1994, p. 65) noted that, 'basing their policies on scientific anthropology, the NAD's definition of "tribe" began to harden into an administrative unit consisting of a chief, a homogenous ethnic group and a piece of land.' Unsurprisingly, the state employed research skills of ethnologists such as Van Warmelo to identify 'tribes' and to map 'tribal' areas within which traditional authorities would be exercised. However, the delimitation of 'tribal' areas was a complex exercise in areas where 'tribes' and ethnic groups were mixed. To that end, the Ministery of Bantu Administration acknowledged that, 'in practice we have now come up against another problem, ... where amongst the Tsonga, for example, there is a farm occupied by the Venda, where the Tsonga occupy another farm in Venda territory' (Union of South Africa, 1959, col. 6698). In such cases, the government decided on the ethnic group and/or tribe that would occupy the designated area, as has been the case in Senwamokgope in the former bantustan of Lebowa (Ramutsindela, 1997).

In practice, the decision meant that people who were found on the wrong side of the ethnic line had to be forcefully removed.

Besides the problem relating to the creation of tribal enclaves, the state had to mediate the positions of traditional authorities as state apparatus and as representatives of designated groups. There were groups of people who had no chief, as evidenced by the communities in Bushbuckridge that Van Warmelo (1935, p. 51) described as,

> a meeting place of tribes from the East, the South, the West, and the North-west ... the result [being] a confused tangle of tribes and sections and scattered units, very often no larger than a family, speaking several languages and following different customs.

The response of the government to such conditions was to manufacture chiefs and to impose them as rulers of certain tribes. The locals who wanted to be chiefs also manipulated the situation, while recalcitrant chiefs were ostracized. For instance, 'Sabata Dalindyebo was deposed as the paramount chief in Thembuland in favour of Kaizer Matanzima' (Hendricks and Ntsebeza, 1999, p. 105).

Subsequently, tribal areas were grouped together to from regions under regional authorities. Vosloo et al (1974) observed that a regional authority consisted of all chiefs and other heads of tribal authorities in the region as ex-officio members. Regional authorities performed 'the general function of advising and making representation to the government on all matters affecting the general interests of blacks within their specific jurisdictions' (South Africa, 2000, p. 10). Two or more regions were combined to form a territory under the control of a territorial authority. That arrangement culminated in the creation of ethnic spaces that formed the cornerstone of bantustans. There, traditional authorities occupied positions as high as that of Chief Ministers. In fact, the so-called paramount chiefs became Chief Ministers in the bantustans and were named presidents in the nominal independent states of Transkei, Bophuthatswana, Venda and Ciskei. Viewed from this angle, chieftainship became

> a public office created by statute. That is the reversal of the position of the chief in traditional society, in which the role of the chief was to represent his people according to the dictates of customary practice. This reversal, effected by the Act, has plainly made the appointment, suspension and deposition of chiefs subject to political manipulation (Khunou, cited in South Africa, 2000, p. 11).

Practically, traditional authorities were placed firmly in local administration under the pretext of Africans taking a greater degree of control of their affairs (Hendricks and Ntsebeza, 1999). Nonetheless, the use of traditional authorities in pursuance of the divide-and-rule policy was later to complicate the status of those authorities in the post-apartheid era.

Traditional authorities in a democratic South Africa

> The introduction of democracy created the need for changes in the structure of African society. Given the needs of modernisation, most of the changes that Africa has been grappling with ever since the advent of democracy have been inescapable. The institution of traditional leadership in South Africa will necessary undergo changes in order to make it more relevant to developing circumstances. Some of these changes may clash with long held values and notions 'sanctified' by history or other factors (South Africa, 2000, p. 7).

There are few, if any, dissenting voices on the need for fundamental changes that are crucial to nurturing a newly found democracy in South Africa. The challenge is how to accommodate institutions such as traditional leaders into a modern democratic system. Chapter 12 of the Constitution of the Republic of South Africa recognizes the institution, status and role of traditional leaders according to customary law (South Africa, 1996). Furthermore, the Constitution makes provision for the incorporation of that institution at various levels of government. For instance, it allows for the establishment of Houses of Traditional Leaders at national and provincial levels. A National House of Traditional Leaders has been established as a statutory body.

> Six provincial houses of traditional leaders were established in terms of the respective provincial acts within the framework of the 1993 Constitution. The number of members of the six provincial houses is as follows: Eastern Cape 20; Free State 15; KwaZulu-Natal 76; Mpumalanga 21; Northern Province 36; and Northwest 24 (South Africa, 2000, p. 43).

Thus, houses of traditional leaders are found in those provinces that inherited a greater portion of former bantustans.

The establishment of houses of traditional leaders after independence/liberation is not a South African phenomenon. Traditional leaders have been accommodated in the House of Assembly in Zimbabwe, the House of Chiefs in Botswana and Zambia, for instance. In South Africa the major problem seems to be the incorporation of traditional leaders at the local level, not least because local areas have been strongholds of chiefs. Thus, the future of traditional leaders is inseparable from transformation at the local government level.

As we have seen in Chapter 5, local government restructuring impacted heavily on local power structures. Notwithstanding contestations between white and black communities who had to share local councils, the demarcation of boundaries created a new context in which traditional leadership could be negotiated. The demarcation took place at two levels: provincial and local. The case of Tafelkop in Northern Province (see Figure 5.2) shows the impact of boundary demarcation on traditional authorities.

Old traditional authority structures in new provinces

Strategies to deal with the legacy of the spatiality of apartheid included the demarcation of new provinces, a task that was carried out by the Commission on the Demarcation/Delimitation of Regions (hereafter CDDR) in 1993 (see Khosa and Muthien, 1998). The CDDR placed the districts of Moutse and Mathanjana of the former KwaNdebele bantustan in Gauteng. The demarcation was disputed by the Ndebele from those districts and other parts of Mpumalanga on the grounds that the provincial boundary divided the Ndebele language group. The Ndebele were divided in opinion over the province to which they should belong. The Pro-Gauteng Co-ordination Committee mobilized mass actions based on Ndebele identity, to demand the inclusion of the former KwaNdebele into Gauteng for economic benefits. This demand was opposed by the Intando Yesizwe Party (IYP) on the grounds that the historical and cultural base of the Ndebele is in Mpumalanga (*Saturday Weekend Argus,* 13/14 January 1996, p. 20). Arguably, the IYP did not want the area to be in Gauteng for fear of competing for seats in a highly-charged political region, as underlined by the following statement from the party: 'we also note that we are the only people with hereditary kingdom placed in a highly urbanised [Gauteng] region' (IYP 9 August 1993, p. 6). The Multi-Party Negotiating Process (MPNP) took a decision to include the former KwaNdebele (Moutse 1, 2 and 3, and Moretele 2) into Mpumalanga for the purposes of the constitution for transition (MPNP, 12 November 1993). The decision to locate KwaNdebele in Mpumalanga satisfied the pro-Mpumalanga group, and was useful to their intention to consolidate the Ndebele in that province (Muthien and Khosa 1995).

Essentially, sections of the Ndebele lived in the former Lebowa bantustan. After 1993, those sections were demarcated into Northern Province – the whole of the former Lebowa has been incorporated into Northern Province. It should be noted that the Ndebele lived among the North Sotho in present-day Northern Province. Implicitly, the issue of provincial preference affected the two ethnic groups, hence there were marked differences among the local people. For instance, the local people were divided between the pro-Northern Province and pro-Mpumalanga camps.

The pro-Northern Province camp had a complex composition because of its political and cultural ramifications. Politically, it had the support of the Northern Province because the (provincial) government wanted Groblersdal to be transferred to the area of jurisdiction of that province in exchange for Bushbuckridge (see Ramutsindela and Simon, 1999). In fact, the (provincial) government told those who did not want to be in Northern Province to 'relocate' to Mpumalanga as suggested by the following statement by the former provincial Minister of Local Government and Traditional Affairs, Mr R.J. Dombo:

> And we have met with those people [pro-Mpumalanga] and indicated to them that they belong to Northern Province. That if they do not want to belong to the Northern province, they can do what in Afrikaans they call: 'vat jou goed en trek Ferreira' [take your properties and go away] (Northern Province 1996, p. 19).

My fieldwork in 1996/97 and 1997/8 showed that the majority (79 per cent) of North Sotho around Groblersdal wanted African settlements to form one TLC with Groblersdal, and that such a TLC be administered by Northern Province. The North Sotho regarded Northern Province as their historical and cultural home, as attested to by the following statements:

> People want to go to Northern Province because they have always been part of Northern Province' (Interview, Mr J. Chabalala, 26 September 1997, Tsantsabela).

> I belong to the North [Northern Province] ... I am a North Sotho (Respondent 24, 20 July 1996, Luckan).

The chairperson of the ANC in Elandskraal, Sam Chauke, was of the view that Groblersdal should be transferred to Northern Province because the town was dominated by North Sotho who belonged to Northern Province. He commented that, 'if you go to Marble Hall and Groblersdal, you will find that there are very few people who come from KwaNdebele' (Interview, 26 September 1997, Elandskraal). The main reason given by the majority (79 per cent) of North Sotho who want to be part of Northern Province is that the former Lebowa bantustan for the North Sotho is in that province (Table 6.1).

Table 6.1 Provincial Preference by Africans around Groblersdal

Ethnic Group	Choice of Province	Percentage
North Sotho	Northern Province	79.06
	Mpumalanga	15.12
	Gauteng	5.81
Ndebele	Northern Province	41.67
	Mpumalanga	45.83
	No Responses	8.33

Source: Field Surveys, 1996-1998

Although the majority of the North Sotho want to be part of Northern Province, Ndebele-speaking residents are split into almost two equal groups on their provincial preference. About 46 per cent of the Ndebele want to join Mpumalanga. The pro-Mpumalanga group feels that there is a need to group the Ndebele in Mpumalanga because their ancestors lived in areas falling under that province, as revealed by the following statements:

> The Ndebele belong to Mpumalanga (Interview, Lesetša Mabona, 25 September1997, Sovenga).

[I want to go to Mpumalanga because] that is where I belong (Respondent 75, 15 July 1996, Groblersdal).

We want to go to [Mpumalanga] as most of our people are from [Mpumalanga] by origin (Mr J.M. Theledi, open letter, *Northern Transvaler,* 7 July 1995, 19).

Undoubtedly, the Ndebele have historical and cultural connections with Mpumalanga. Roossenekal, in the Middelburg district of Mpumalanga is considered the historical headquarters of the Ndebele. The IYP described the importance of Roossenekal to the Ndebele in these words:

On the 19th of December of each year, the Ndebele converge from Standerton, Middelburg, Groblersdal, Bronkhorstspruit, Delmas, Bethal, Pretoria as well as KwaNdebele to Roossenekal where they hold a cultural feast and a historical commemoration of their ancestral leaders whose remains are buried there and whose spirits are still there (IYP 9 August 1993).

Historical connections aside, the IYP evoked Ndebele history in order to mobilize political support along ethnic lines. Those Ndebele in Northern Province demanded reincorporation into Mpumalanga in order to strengthen their constituency there. However not all the Ndebele in the vicinity of Groblersdal supported reincorporation into Mpumalanga. A significant proportion (42 per cent) wanted to remain in Northern Province, for various reasons, including the political and economic ties they have established in that province.

For their part, the North Sotho in Tafelkop were divided over their preference for provinces. The two chiefs in Tafelkop, Matsepe and Rammupudu, preferred Northern Province and Mpumalanga respectively. A section under chief Rammupudu played a significant role in the boundary dispute, mainly because it did not consider itself to be of North Sotho origin – hence it felt that it had no cultural connections to Northern Province.

Rammupudu's subjects (the Kopa) have Tswana roots. Mokgokong (1966) noted that their original home is Barolong around the present town of Mafikeng (North West). Their presence in Maleoskop near Groblersdal resulted mainly from migration. They were 'cleared' from Maleoskop in 1962 and were resettled in the Nebo district of the former Lebowa bantustan. Their chief, Rammupudu, served in (but was later fired from) the cabinet of Chief Minister Ramodike of Lebowa in the 1980s (Interview, David Tshehla, 29 October 1997, Monsterlus).

Arguably, the loss of a parliamentary seat by chief Rammupudu in the former Lebowa served to create the anti-Northern Province feeling shared by some of his followers. The anti-Northern Province perception was clear from (pro-Mpumalanga) respondent no. 29, who commented that, 'Northern Province cannot do anything for us.' Perceptions of this nature have their roots in the bantustan politics of patronage that marginalized areas in which opposition politicians resided. Those areas did not benefit from governmental projects (if there were any), and residents could hardly be employed in the civil service. It is against that background that Rammupudu and his followers did not want to be in Northern Province from where they lost favour in the former Lebowa – a continuation of bantustan stereotypes. Unsurprisingly, Chief Rammupudu and Prince James Mahlangu of KwaNdebele wanted the creation of a 10th region, failing of which they would opt for reincorporation into Mpumalanga (Interview, David Tshehla, 29 October 1997, Monsterlus).

The pro-Mpumalanga camp exhibited other complexities. Chief Rammupudu successfully lodged a land claim in Maleoskop (Mpumalanga) near Groblersdal. The South African National Defence Force that used Maleoskop as a training base has agreed that land ownership of the area should be restored to the Bakgaga (Kopa), but would negotiate the use of the land for training (South African Broadcasting Corporation, 19 February 1998). Although the land claim has been used to regain land ownership of Maleoskop, it also serves to support Rammupudu's claim that he and his subjects belong to Mpumalanga. The ultimate aim of Chief Rammupudu is to consolidate his chieftaincy in Mpumalanga, away from his rival, Chief Matsepe, in Tafelkop (Northern Province). As Leonard Malatsi said,

> Rammupudu's people cannot be scattered into two provinces, hence the claim to belong to Mpumalanga is also an attempt to consolidate Rammupudu's clan, i.e. [to] bring [the] people of Tafelkop and Dennilton [Mpumalanga] together (Interview, Leonard Malatsi, 29 October 1997, Groblersdal).

There were also claims of a split in the ANC ranks on the issue of allocating the Tafelkop/ Motetema/Groblersdal area to a particular province (*Sowetan* 9 May 1996, p. 4; Interview, Vickus Coetser, 25 September 1997, Pietersburg). Thus, the provincial preferences in these areas assume racial, ethnic and political dimensions. More crucially, the demarcation of provincial boundaries opened up the politics of traditional leadership that had been suspended in the apartheid era. That politics is more visible in contested local government boundaries.

Traditional leaders and the redemarcation of municipalities

While local government elections in 1995/6 might have sealed democracy at the local level in South Africa, the establishment of local councils was certainly a worrying aspect to many traditional leaders. For instance, Chief Ncamashe in the Eastern Cape regarded interim local government structures as reducing his role to an administrator of petty customary laws (*Sunday Times*, 5 March 1995, p. 10). These sentiments became more pronounced in the redemarcation of municipalities in 1999/2000. As I have intimated above, the LGTA entrenched the rural-urban dichotomy through the classification of rural and urban local councils. Consequently, most rural areas had neither capacity nor resources for development. The redemarcation of municipalities in 1999/2000 is set to address these problems. Of significance in the redemarcation process are efforts to amalgamate rural and urban areas into common municipalities where this is feasible. That process took place against the lack of clarity on the status of traditional leaders at the local level. This was not properly addressed in the transition because it was probably being viewed by dominant urban-based politicians as less critical to the transformation of South Africa. There is a well-established perception of traditional leaders as colonial and/or apartheid structures (*Dispatch Online*, 10 March 2000). Thus, they are viewed as irrelevant to a democratic system and its development imperatives. There are doubts on whether these leaders have the capacity to improve the lives of rural dwellers. More importantly, their legitimacy is also being seriously questioned by proponents of (western) democracy – they are seen as making nonsense of democracy.

Critics of the institution of traditional leaders are also concerned about the financial cost of maintaining that institution. For example, the six paramount chiefs in the Eastern Cape received R 322 800 a year, while the 221 chiefs there each received R77 472 a year from April 1999 (*Electronic Mail & Guardian*, 21 April 1999). Critics view this as a waste of the tax-payer's money on unelected leaders.

For their part traditional leaders see the re-demarcation of municipalities as a process that 'will incorporate tribal land and leave them with no power or influence' (*Sowetan*, 27 October 1999, p. 2). What is instructive about the renewed concerns over municipalities by chiefs, is how the entrenched power of chiefs in the bantustans continues to haunt changes in post-apartheid South Africa. In Kwa-Zulu Natal, the historical construction of chieftaincy in the bantustan and the association of that institution with the Inkatha Freedom Party make the collapsing of 'tribal' areas there an archilles heel. Unsurprisingly, the Amakhosi (chiefs) in that area are opposed to the introduction of municipal structures in rural areas (*Sowetan*, 26 January 2000, p. 3), because such new structures are seen to be interfering with the boundaries of 'tribal' land and the powers of chiefs. The President of the Congress of Traditional Leaders of South Africa (CONTRALESA), *khosi* Patekile Holomisa, pointed out that 'traditional authorities must be recognised as the primary form of rural local government' (*Mail & Guardian*, 11-17 February 2000b, p. 29). He further maintains that,

> the government and the Municipal Boundaries Demarcation Board must at all costs ensure that no traditional authority territorial area is divided between more than one municipal council. Should such division occur it will be seen as a diminution of the tribal land concerned and an addition to another traditional authority territorial area. We take as nonsense the suggestion that municipal boundaries will not affect the jurisdiction of traditional authorities even where they straddle tribal authorities (*Mail & Guardian*, 11-17 February 2000b, p. 29).

The words of King Justice Mpondmbini Sigcau of the Pondo reveal the same position on chieftaincy and 'tribal' areas:

> The whites told us what is to be done. Now we are faced with the same old bogey again. Now a government department is telling us what is to be done. Pondos are very sensitive about local loyalties; who belongs where and with whom. Now you want to curve us up in the same way as the slave states came into being? (*Mail & Guardian*, 4-10 February 2000a, p. 37).

At face value, the issue might appear to be safeguarding 'traditional authority territorial areas'. However, the implication of the concerns of chiefs goes deeper into the interface of the 'traditional' and modern rule, and the very essence of the direction and purpose of transformation. The system of traditional authorities seems to have socialized people into space and the powers that go with it. The implication of opposing the redemarcation is that the core areas of bantustans and the powers of chiefs should be maintained and retained. Hendricks and Ntsebeza (cited in *Mail & Guardian*, 18 to 24 February 2000c, p. 33) consider the retention of 'tribal' areas and their chiefs in post-apartheid South Africa as a 'return to the homeland system preferably with a form of Bantu Authorities in charge of local government.'

Conclusion: the future of the past

As much as we cannot go back completely to pre-colonial times, we cannot rush into modernity by undermining and discarding our past. A proper role of traditional leaders at the local level has to be determined. The White Paper on Local Government foresees a co-operative model for rural local governance. The Constitution provides for a role for traditional leadership at local level (South Africa, 2000, p. 13).

In terms of local government restructuring, the constitution of South Africa has not offered specific guidelines on changing the rural-urban duality. Instead, the Interim Constitution perpetuated that division by drawing the distinction between urban and rural local councils. As I have shown above, the demarcation of new local councils represents attempts to transcend that divide. Notwithstanding such attempts, there has been lack of clarity of the position and roles of traditional leaders throughout the transitional period (1994-1999). This has been acknowledged by the White Paper on Local Government (South Africa, 1998). However, attempts to launch the final phase of local government in 1999 placed the issue of traditional leaders firmly on the agenda for transformation.

Traditional leaders want to maintain their institution in the new political dispensation. They argue for respect of traditional authority areas and protest the mixing or splitting of their areas through the redemarcation of municipalities. At issue are the roles, functions and powers of traditional leaders. These have not been clarified in the constitution. A belated attempt to revisit the issue of traditional leaders came in April 2000 in the form of a discussion document on traditional leaders. While the discussion document could pave the way for resolving the issue of traditional leaders, the redemarcation process pre-empts the future of areas under traditional leaders. Moreover, other processes such as land reform also impinge upon these areas. There is therefore uncertainties and well-established fears among traditional leaders, hence those leaders threatened to derail the local government elections of 5 December 2000.

References

Bekker, J.C. (1993), 'The role of chiefs in a future South African constitutional dispensation', *Africa Insight*, vol.23, pp. 200–204.

Dispatch Online. (2000), 10 March, East London.

Electronic Mail & Guardian. (1999), 21 April, Johannesburg.

Hendricks, F. and Ntsebeza, L. (1999), 'Chiefs and rural local government in post-apartheid South Africa', *African Journal of Political Science*, vol.4, pp. 99–126.

Intando Yesizwe Party. (1993), *Presentation by Intando Yesizwe Party during the debate on the Report of the Demarcation of Regions*, 9 August, Kempton Park.

Khosa, M.M. and Muthien, Y.G. (1998), *Regionalism in the new South Africa*, Ashgate, Aldershot.

Mail & Guardian. (2000a), 4-10 February, p. 37, Johannesburg.

Mail & Guardian. (2000b), 11-17 February, p. 29, Johannesburg.

Mail & Guardian. (2000c), 18-24 February, p. 33, Johannesburg.

Mamdani, M. (1996), *Citizen and Subject: contemporary Africa and the legacy of late colonialism*, Princeton University Press, Princeton, N.J.

Muthien, Y.G. and Khosa, M.M. (1995), 'The kingdom, the volkstaat and the new South Africa: drawing South Africa's new regional boundaries', *Journal of Southern African Studies*, vol. 21, pp. 303-22.

Northern Province. (1996), *Hansard*, Pietersburg.

Northern Transvaler. (1995), 7 July, p. 19, Pietersburg.

Ramutsindela, M.F. (1997), 'The making of territories: a case study in South Africa' In A. Awotona and N. Teymur (eds), *Tradition, Location and Community: place-making and development*, Avebury, Aldershot, pp. 225–33.

Ramutsindela, M.F. and Simon, D. (1999), 'The politics of territory and place in post-apartheid South Africa: the disputed area of Bushbuckridge', *Journal of Southern African Studies*, vol. 25, pp. 479-98.

Ritchken, E. (1994), *Leadership and Conflict in Bushbuckridge: struggles to define moral economies within the context of rapidly transforming political economies (1978-1990)* (PhD Thesis), University of the Witwatersrand, Johannesburg.

Saturday Weekend Argus. (1996), 13/14 January, p. 20, Cape Town.

South Africa. (1996) *Constitution of the Republic of South Africa*, Government Printer, Pretoria.

South Africa. (1998), *White Paper on Local Government*, Government Printer, Pretoria.

South Africa. (2000), *Draft discussion document towards a White Paper on Traditional Leadership and Institutions*, Pretoria, Department of Provincial and Local Government.

South African Broadcasting Corporation. (1998), 19 February, Auckland Park.

Sowetan. (1996), 9 May, p. 4, Johannesburg.

Sowetan. (1999), 27 October, p. 2, Johannesburg.

Sowetan. (2000), 26 January, p. 3, Johannesburg.

Sunday Times. (1995), 5 March, p. 10, Johannesburg.

Union of South Africa. (1959), *Hansard*, Pretoria

Van Warmelo, N.J. (1935), *A preliminary survey of the bantu tribes of South Africa*, Government Printer, Pretoria.

Vosloo, W.B. et al. (1974), *Local government in Southern Africa*, Academia, Pretoria.

7 Conclusion: Into the Future

We are on course. Steadily the dark clouds of despair are lifting, giving way to our season of hope. Our country which, for centuries, has bled from a thousand wounds is progressing towards its healing. The continuing process of social and national emancipation, to which we are all subject, constitutes an evolving act of self-definition. At the dawn of a new life, our practical actions must ensure that none can challenge us when we say – we are a nation at work to build a better life! (Mbeki, 25 June 1999).

The analysis presented in this book has shown that the transition to a post-apartheid state in South Africa should be understood in the context of the nature of the post-independence state in Africa. In other words, the analysis of the post-apartheid state requires theories (of the state) that recognize the nature of the state at hand. Such has been a demand in the understanding of the post-colonial state in Africa. As we have seen, the post-colonial state cannot be appropriately studied through the lenses of Western theories of the state; to do so would lead to the mischaracterization of the state in Africa. Various Western theories of the state, ranging from liberal to Marxist paradigms, assume the existence of a society and/or class structure that are almost non-existent or at most fragmentary in Africa. The salient features of African states have not only been ignored by conventional Western theories of the state, but also elude the now fashionable discourses of globalization. Despite the contested meanings and implications of globalization, there has not been a clear differentiation between the impact of that process in countries of the South and the North. Instead, scholars tend to treat globalization and the state as if the state were a homogenous entity. This begs the question of how the post-apartheid state should be approached.

As I have argued in this book, the reconstruction of the post-apartheid state should be understood in the context of the post-colonial state in Africa. South Africa has to confront some of the very thorny problems that faced post-independence Africa. These include the national question (Chapter 3), the land question (Chapter 4) and issues of sub-national territorial reorganization (Chapters 5 and 6).

South Africa faces all these challenges against the background of pessimistic scenarios that unfolded in post-independence Africa. Problems around national identities, for instance, have not been resolved in most African states for reasons discussed in Chapters 2 and 3. As I have shown in Chapter 3, the analysis of nation-building in South Africa should be seen in the context of that process in Africa and beyond. Evidently, there are nation-states in the South and the North that had and still have to deal with problems of national identities. The process of nation-building and the results therefrom are unpredictable, more especially because nation-building is contingent on approaches and the dynamics of centrifugal and centripetal forces in operation in different countries. Moreover, such 'projects' are often not stable and irreversible as violent fragmentation in the former USSR and Yugoslavia has shown all too clearly in the course of the post-Cold War geopolitical transitions. In South Africa, the approach to nation-building has been to seek unity in diversity in a shared state. For the moment, that approach has brought stability to a hitherto deeply divided society.

The mechanisms for addressing the legacy of apartheid, though, could revive divisions in that society. As we have seen, racial divisions are still very much strong. Those divisions are likely to persist because issues of race and interests in South Africa run parallel to those of race and ethnicity (Mates, 1995). Policies such as affirmative action that aim to correct historical imbalances by helping disadvantaged communities (i.e. blacks) are misconstrued by sections of the white community as apartheid in reverse. Furthermore, opponents of that policy see it as leading to the rise of African hegemony. Practically, most of the strategies to address the legacy of apartheid will touch the race question, all the more because the apartheid state thrived on racial and ethnic categorization. For instance, land ownership patterns reflected those categorizations.

It follows that the land question in South Africa was/is unavoidable because of the historically racially skewed distribution of land, the political, economic and social meanings and implications of land ownership and dispossession, and the now-dominant post-apartheid vision of a democratic order. All these, required that the land question be discussed during the negotiations in the early 1990s. The results were the adoption of a market-driven land reform process, as I explained in Chapter 3. Land reform is driven through three programmes, namely, land redistribution, land tenure reform and restitution. In all these, the state mediates between victims and beneficiaries of land dispossession in order to achieve the objectives of land reform and to maintain national stability.

In the Makuleke land claim, the nationally driven land reform process provided the context within which the Makuleke were able to regain their land rights in Pafuri. The Makuleke claim represents one of the first large and community-based claims to be resolved, thus serving as a model for land claims in 'sensitive' areas. In that claim, it was possible to negotiate conflicting interests between conservation and the political imperative of restitution. The restoration of land rights was achieved within the ambit of national reconstruction, but the nature of restoration was exceptional to the general tenets of restitution (see Chapter 4). 'Land rights' had to be redefined to accommodate conflicting national and community interests. In this context, the application of a flexible land reform in South Africa is likely to create different avenues for addressing the land question. What is also instructive about the Makuleke case is how the local community used land reform as a platform on which to restore its identity. This was never conceived as the aim of restitution. As we have seen, the Makuleke were able to reassert their status that was lost through apartheid social engineering. The land claim also enabled them reclaim their chieftaincy from Chief Mhinga, to whom they had been subjected since their forced removal in 1969.

The urgent need for land reform in South Africa is an indication of the problem of spatiality that the new state has inherited. It is against this background that the new state not only focused on land reform, but also embarked on the process of re-drawing internal boundaries. As shown in Chapters 5 and 6, apartheid boundaries were/are incompatible with the vision of a non-racial, democratic South Africa. Unsurprisingly, the re-drawing of provincial boundaries ensued in 1993 as an attempt to overcome racial and ethnic divisions in that country. Attempts to achieve the same goals on a smaller scale were made through the re-demarcation of local government boundaries in 1994/5. All these efforts are bound up with the broader aim of reconstructing the post-apartheid state to achieve a non-racial political dispensation.

That process was highly contentious on the ground, more especially because constitutional measures privileged the local arena. Local communities were to negotiate the process of integration. As shown in Chapter 5, even the progressive areas such as Johannesburg had difficulties in establishing metropolitan government structures. In that city, communities were divided over the establishment of a single metropolitan government (the megacity). The official view is that metropolitan government creates a basis for equitable and socially just metropolitan governance (South Africa, 1998). However, the mechanisms for achieving that equity generated much-heated debate.

If progressive areas are transformed with difficulties, what more of conservative towns such as Groblersdal. The white community in that town opposed the integration of black and white people into a common local council. It employed the opportunities created by the process of local government restructuring to preserve and perpetuate apartheid boundaries. That aim was vehemently opposed by the government, which saw Groblersdal as a white enclave. In other words, the purpose of the white community of Groblersdal was against the spirit and intent of the new political dispensation. Local government restructuring in that area failed dismally, and could not even get off the ground during the entire transition period. The main factors behind that failure are: a strong resistance by the white community of Groblersdal to amalgamation with surrounding African areas, the anti-Northern Province stance of the white and sections of the African community there, the roles played by politicians, and loopholes in the Local Government Transition Act of 1993.

Thus, the de-racialization of the local state in Groblersdal did not succeed during the transition period (1994-1999), making Groblersdal the only apartheid municipality in the country. The failure of restructuring there is partly blamed on conflict of interests among the African communities adjacent to Groblersdal. Those communities are divided along ethnic lines over the choice of province. That is, in Groblersdal there is the (re)emergence of racial, ethnic and political alliances that define the axis of local politics.

In all accounts the reconfiguration of apartheid spaces provided the context for local articulations. Historical constructions such as the institution of traditional leaders had to be negotiated. Their oppressive roles in colonial and apartheid years aside, traditional leaders cannot be wished away. For the moment, the institution of traditional leaders has been incorporated at national and provincial levels. The difficulty, though, is in accommodating it at the local level. There has been lack of clarity on the status of traditional leaders in the new dispensation. Unsurprisingly, chiefs contested the launching of the final phase of local government. At the time of writing (i.e. October 2000), efforts were being made to diffuse the stand off between traditional leaders and the government.

This book as a whole has attempted to show the pathways of post-apartheid transformation and to provide the link between various threads of national reconstruction and to show how these are contested on the ground.

It is hoped that this volume will contribute towards understanding the challenges of post-independence reconstruction.

References

Mates, R. (1995), *The Election Book: judgements and choices in South Africa's 1994 election*, Institute for Democracy in South Africa, Cape Town.

Mbeki, T. (1999), *Address at the Opening of Parliament, National Assembly*, 25 June, Cape Town.

South Africa, (1998), *White Paper on Local Government*, Government Printer, Pretoria.

Select Bibliography

African National Congress. (1994), *The Reconstruction and Development Programme*, Umanyano, Johannesburg.

Agnew, J. (1999), 'The new geopolitics of power', In D. Massey, J. Allen and P. Sarre (eds), *Human Geography Today*, Polity Press, Cambridge, pp. 173–193.

Ake, C. (1996), *Democracy and Development in Africa*, Bookings Institution, Washington, D.C.

Amin, S. (1993), 'South Africa in the global economic system', *Work in Progress*, vol. 87, pp. 10–11.

Asiwaju, A.I. (1996), 'Borderlands in Africa: a comparative research perspective with particular reference to Western Europe', In P. Nugent and A.I. Asiwaju (eds), *African Boundaries: barriers, conduits and opportunities*, Pinter, London, pp. 253–65.

Bayart, J-F, (1993), *The State in Africa: the politics of the belly*, Longman, London.

Beinart, W. (1994), *Twentieth-century South Africa*, Opus, London.

Bekker, J.C. (1993), 'The role of chiefs in a future South African constitutional dispensation', *Africa Insight*, vol.23, pp. 200–204.

Benson, M. (1985), *South Africa: struggle for a birthright*, International Defence and Aid Fund for Southern Africa, London.

Berman, B.J. (1998), 'Ethnicity, patronage and the African state: the politics of uncivil nationalism', *African Affairs*, vol. 91, pp. 305–41.

Bickford-Smith, V. (1989), 'A special tradition of multi-racialism'? segregation in Cape Town in the late nineteenth and early twentieth centuries', In G.J. Wilmont, and M. Simons, (eds), *The angry divide: social and economic history of the Western Cape*, David Philip, Cape Town, pp. 47–62.

Binswanger, H.P. and Deininger, K. (1993), 'South African land policy: the legacy of history and current options', *World Development*, vol. 21, pp. 1451–75.

Blumenfeld, J. (1997), 'From icon to scapegoat: South Africa's Reconstruction and Development Programme', *Development Policy Review*, vol. 15, pp.65–91.

Blumenfeld, J. (1999), 'The post-apartheid economy: achievements, problems and prospects', In J.E. Spence (ed.) *After Mandela: the 1999 South Africa elections*, Royal Institute of International Affairs, London.

Boateng, E.A. (1978), *A Political Geography of Africa*, Cambridge University Press, Cambridge.

Brown, L. (1998), 'WESSA's concerns over Makuleke claim addressed', *African Wildlife*, vol. 52, p. 9.

Cameron, R. (1995), 'The history of devolution of powers to local authorities in South Africa: the shifting sands of the state control', *Local Government Studies*, vol.21, pp. 396–417.

Carroll, B.W. and Carroll, T. (1997), 'State and ethnicity in Botswana and Mauritius: a democratic route to development', *Journal of Development Studies*, vol. 33, pp. 464–486.

Carruthers, J. (1995), *The Kruger National Park: A social and political history*. University of Natal Press, Pietermaritzburg.

Chabal, P (ed.). (1986), Political Domination in Africa: reflections on the limit of power, Cambridge University Press, New York.

Christopher, A.J. (1994), 'Indigenous land claims in the Anglophone world', *Land Use Policy*, vol. 11, pp. 31–44.

Clapman, C. (1999), 'Sovereignty and the Third World state', *Political Studies*, vol. 47, pp. 522–537.

Coleman, J.S. (1994), *Nationalism and Development in Africa*, University of California, Berkeley.

Cooper, K. (1998), 'Editorial', *African Wildlife*, vol. 52, p. 7.

Davidson, B. (1981), *The People's Cause: a history of guerrillas in Africa*, Longman, London.

Davidson, B. (1990), 'The crisis of the nation-state in Africa', (interviewed by Munslow), *Review of African Political Economy*, vol. 49, pp. 9–21.

De Kiewiet, C.W. (1941), *A history of South Africa: social and economic*, Oxford University Press, London.

De Villiers, B. (1999), *Land claims and national parks: the Makuleke experience*. Pretoria: Human Sciences Research Council.

Dewar, D. (1995), 'The urban question in South Africa: the need for a planning paradigm shift', *Third World Planning Review*, vol. 17, pp. 407–418.

De Wet, C. (1997), 'Land reform in South Africa: A vehicle for justice and reconciliation, or a source of further inequality and conflict?' *Development Southern Africa*, vol. 14, pp. 355–62.

Dicke, B.H. (1926), 'The first voortrekkers to the Northern Transvaal and the massacre of the Van Rensburg Trek', *South African Journal of Science*, vol. 23, pp. 1006-1021.

Doornbos, M. (1990), 'The African state in academic debate: retrospect and prospect', *Journal of Modern African Studies*, vol. 28, pp. 179–198.

Doornbos, M. and Markakis, J. (1994), 'Society and State in Crisis: what went wrong in Somalia'? *Review of African Political Economy*, vol. 59, pp. 82–89.

Dubow, S. (1994), 'Ethnic euphemisms and racial echoes', *Journal of Southern African Studies*, vol. 30, pp. 355–70.

Ergas, Z. (1987), 'Introduction', In Z. Ergas (ed.), *The African State in Transition*, Macmillan, London, pp.1–22.

February, V. (1991), *The Afrikaners of South Africa*, Kegan Paul, London.

Fisch, J. (1988), 'Africa as terra nullius: the Berlin Conference and International Law', in S. Forster, W.J. Mommsen and R. Robinson (eds), *Bismarck, Europe and Africa: The Berlin Africa Conference 1884-1885 and the onset of partition*, Oxford University Press, Oxford, pp. 347–375.

Gakwandi, A.S. (1996), 'Towards a new political map of Africa' in T.Abdul-Raheem (ed.), *Pan Africanism: politics, economy and social change in the twenty-first century*, Pluto, London, pp. 181–190.

Gellner, E. (1983), *Nations and nationalism*, Blackwell, Oxford.
Goldin, I. (1987), *Making race: the politics of economics and coloured identity in South Africa*, Longman, London.

Guelke, A. (1996), 'Dissecting the South African miracle: African parallels', *Nationalism and Ethnic Politics*, vol. 2, pp. 141-54.

Hangula, L. (1993), *The International Boundary of Namibia*. Gamsberg-Macmillan, Windhoek.

Hargreaves, J.D. (1988), 'The Berlin Conference, West African boundaries and the eventual partition', in S. Forster, W.J. Mommsen and R. Robinson (eds) *Bismarck, Europe and Africa: The Berlin Africa Conference 1884-1885 and the onset of partition*, Oxford University Press pp. 313–20.

Harries, P. (1987), 'A forgotten corner of the Transvaal': reconstructing the history of a relocated community through oral testimony and song' In B. Bozzoli (ed.), *Class, community and conflict: southern African perspectives*, Ravan Press, Johannesburg, pp. 93–134.

Harvey, D. (1996), 'On planning the ideology of planning', In S. Campbell and S. Fairnstein (eds), *Reading in Planning Theory*, Blackwell, Oxford, pp. 176–97.

Hendricks, F. and Ntsebeza, L. (1999), 'Chiefs and rural local government in post-apartheid South Africa', *African Journal of Political Science*, vol.4, pp. 99–126.

Herbst, J. (2000), States and Power in Africa: comparative lessons in authority and control, Princeton University Press, Princeton, N.J.

Howell, J.M. (1994), Understanding Eastern Europe, Kogan Page, London.

Humphries, R. (1991), 'Wither Regional Services Council', In M. Swirling, R. Humphries and S. Khehla (eds), *Apartheid city in transition*, Oxford University Press, Cape Town, pp. 78–90.

Jones, P.S. (1999), '"To come together for progress": modernization and nation-building in South Africa's bantustan periphery – the case of Bophuthatswana', *Journal of Southern African Studies*, vol. 25, 579–605.

Johnson, N.C. (1992), 'Nation-building, language and education: the geography of teacher recruitment in Ireland, 1925-55', *Political Geography*, vol. 11, pp. 170–189.

Kamrava, M. (1993), 'Conceptualising Third World politics: the state-society see-saw', *Third World Quarterly*, vol. 14, pp. 703–716.

Katzellenbogen, S. (1996), 'It didn't happen at Berlin: politics, economics and ignorance in the setting of Africa's colonial boundaries', In P. Nugent and A.I. Asiwaju (eds), *African Boundaries: barriers, conduits and opportunities*, Pinter, London, pp. 21–34.

Keith, M. and Cross, M. (1993), 'Racism and the postmodern city', In M. Cross and M. Keith (eds), *Racism, the city and the state*, Routledge, London, pp. 1–30.

Khosa, M. M. and Muthien, Y.G. (eds). (1998), *Regionalism in the New South Africa*, Ashgate, Aldershot.

Knight, D.B. (1985), Territory and people or people and territory? Thoughts on postcolonial self-determination. *International Political Science Review*, vol. 6, pp. 48–272.

Lemon, A. (1996), 'The new political geography of the local state in South Africa', *Malaysian Journal of Tropical Geography*, vol. 27, pp. 35–45.

Levin, R. and Weiner, D. (1996), 'The politics of land reform in South Africa after apartheid: perspectives, problems and prospects', *Journal of Peasant Studies*, vol. 23, pp. 93–119.

Levin, R. and Weiner, D (eds). (1997), *No More Tears: struggles for land in Mpumalanga, South Africa, Africa,* World Press, Trenton, NJ.

Lonsdale, J. (1992), 'The conquest state of Kenya, 1895-1905', In B. Berman and J. Lonsdale, *Unhappy valley (Book Two),* James Currey, London, pp. 13–44.

Mabin, A. (1995), 'On the problems and prospects of overcoming segregation and fragmentation in Southern African cities in the postmodern era', In S. Watson and K. Gibson (eds) *Postmodern cities and space*, Blackwell, Oxford, pp. 187–98.

Mabin, A. (1999), 'From hard top to soft serve: demarcation of metropolitan government in Johannesburg', In R. Cameron, *A Tale of three cities*, Van Schaik, Pretoria, pp. 159–200.

Maharaj, B. (1997), 'The politics of local government restructuring and apartheid transformation in South Africa', *Journal of Contemporary African Studies*, vol. 15, pp. 261–285.

Maharaj, G (ed). (1999), *Between Unity and Diversity: essays on nation-building in post-apartheid South Africa*, David Philip, Claremont.

Mamdani, M. (1996), *Citizen and Subject: contemporary Africa and the legacy of late colonialism*, Princetown University Press, Princeton, NJ.

Manzo, K.A. (1996), *Creating boundaries: the politics of race and nation*, Lunne Rienner, Boulder.

Marais, H. (1998), *South Africa – Limits to Change: the political economy of transformation*, Zed Books, London.

Mare, G. (1993), *Ethnicity and Politics in South Africa*, Zed Books, London.

Marks, S. (1986), *The ambiguities of dependence in South Africa: class, nationalism and the state in twentieth century Natal*, John Hopkins University Press, Baltimore.

Marks, S. and Trapido, S. (1987), 'The politics of race, class and nationalism', In S. Marks and S. Trapido (eds), *The politics of race, class and nationalism in twentieth century South Africa*, Longman, London, pp. 1–70.

Mason, J.W. (1996), *The Cold War: 1945-1991*, Routledge, London.

Mates, R. (1995), *The Election Book: judgements and choices in South Africa's 1994 election*, Institute for Democracy in South Africa, Cape Town.

Mayall, J. (1992), 'Nationalism and international security after the Cold War', *Survival*, pp. 19–35.

Mazrui, A. (1980), *The African Condition: a political diagnosis*, Heinemann, London.

Mazrui, A. and Tidy. M. (1984), *Nationalism and the new states in Africa*, Heinemann, London.

Mbembe, A. (1992), 'Provisional Notes on the Postcolony', *Africa*, vol. 62, pp. 3–37.

Miles, W.F.S. (1995), 'Decolonization and disintegration: the disestablishment of the state in Chad', *Journal of Asian and African Studies*, vol. 30, pp. 41–52.

Munslow, B. and FitzGerald, P. (1997), 'The search for a development strategy: the RDP and beyond', In P. FitzGerald, A. McLennan and B. Munslow (eds), *Managing Sustainable Development in South Africa*, Oxford, Cape Town, pp. 41–61.

Muthien, Y.G. and Khosa, M.M. (1995), 'The kingdom, the volkstaat and the new South Africa: drawing South Africa's new regional boundaries', *Journal of Southern African Studies*, vol. 21, pp. 303-22.

Nel, M. (1996), 'Kruger Land Claim: South Africa's premier national park faces a major challenge', *African Wildlife*, vol. 50, pp. 6–9.

Nkrumah, K. (1961), *I Speak of Freedom*, Panaf Books, London.

Nyerere, M.J. (1968), *Nyerere: Freedom and Socialism*, Oxford University Press, Nairobi.

O'Laughlin, B. (1995), 'Past and present options: land reform in Mozambique', *Review of African Political Economy*, vol. 63, pp. 99–106.

O'Meara, D. (1983), *Volkskapitalisme: class, capital and ideology in the development of Afrikaner nationalism*, 1934-1948, Cambridge University Press, Cambridge.

Paasi, A. (1996), Territories, Boundaries and Consciousness: the changing geographies of the Finish-Russian border, Wiley, New York.

Parnell, S. (1997), 'South Africa's cities: perspectives from the ivory tower of urban studies', *Urban Studies*, vol. 34, pp. 891-906.

Ramutsindela, M.F. (1997), 'The making of territories: a case study in South Africa' In A. Awotona and N. Teymur (eds), *Tradition, Location and Community: place-making and development*, Avebury, Aldershot, pp. 225–33.

Ramutsindela, M.F. (1998), 'The survival of apartheid's last town council in Groblerdal, South Africa', *Development Southern Africa*, vol. 15, pp. 1–12.

Ramutsindela, M.F. (2000), 'African boundaries and their interpreters', In N. Kliot and D. Newman (eds), Geopolitics at the end of the twentieth century: the changing world political map, Frank Cass, London, pp. 180–198.

Ramutsindela, M.F. (2001), 'Down the post-colonial road: reconstructing the post-apartheid state in South Africa', *Political Geography*, vol. 20, pp. 57-84.

Ramutsindela, M.F. and Simon, D. (1999), 'The politics of territory and place in post-apartheid South Africa: the disputed area of Bushbuckridge', *Journal of Southern African Studies*, vol. 25, pp. 479-98.

Ramsamy, E. (1996), 'Post-settlement South Africa and the national question: the case of the Indian minority', *Critical Sociology*, vol. 22, pp. 57-77

Ranger, T. (1983), 'The invention of tradition in colonial Africa', In E. Hobsbawn and T. Ranger (eds), *The Invention of Tradition*, Cambridge University Press, Cambridge, pp. 211–262.

Seethal, C. (1991), 'Restructuring the local state in South Africa: Regional Services Councils and crisis resolution', *Political Geography Quarterly*, vol. 10, pp. 8–25.

Sidaway, J.D. and Simon, D. (1993), 'Geopolitical transition and state formation: the changing political geographies of Angola, Mozambique and Namibia', *Journal of Southern African Studies*, vol. 19, pp. 6–28.

Simon, D. (1993), 'The communal land question revisited', *Third World Planning Review*, vol. 15, pp. iii–vii.

Simon, D. (1994), 'Putting South Africa(n geography) back into Africa', *Area*, vol. 26, pp. 296–300.

Simon, D. (1996), 'Restructuring the local state in post-apartheid cities: Namibian experience and lessons for South Africa', *African Affairs*, vol. 95, pp. 51–84.

Simon, D. and Ramutsindela, M.F. (2000), 'Political geographies of change in southern Africa', In R.C. Fox and K.M. Rowntree (eds) *The Geography of South Africa in a Changing World*, Oxford University Press, Cape Town, pp. 89–113.

Simpson, M. (1994), 'The experience of nation-building: some lessons for South Africa', *Journal of Southern African Studies*, vol. 20, pp. 463–74.

Short, J.R. (1993), *An introduction to political geography*, Routledge, London.

Smit, D.W.J. (1996) 'Leave this "diamond" unmined', *Conserva*, vol. 12, pp. 11–13.

Smith, A.D. (1976) 'Introduction', In A.D. Smith (ed.), *National movements*, Macmillan, London, pp. 1–30.

South Africa. (1991), *Minerals Act*, Government Printer, Pretoria.

South Africa. (1983), *Constitution of the Republic of South Africa*, Government Printer, Pretoria.

South Africa. (1993), *Local Government Transition Act*, Government Printer, Pretoria.

South Africa. (1995), *Communal Association Bill*, Government Printer, Pretoria.

South Africa. (1996), *Constitution of the Republic of South Africa*, Government Printer, Pretoria.

South Africa. (1998a), *Municipal Demarcation Act*, Government Printer, Pretoria.

South Africa. (1998b), *White Paper on Local Government*, Government Printer, Pretoria.

South Africa. (2000), *Draft discussion document towards a White Paper on Traditional Leadership and Institutions*, Pretoria, Department of Provincial and Local Government.

South African Institute of Race Relations. (1995), 'Freehold key to land reform', *Frontiers of Freedom*, vol. 5, pp. 7–9.

South African Institute of Race Relations. (1997), 'Give the people their land', *Frontiers of Freedom*, vol. 12, pp. 18–20.

Stark, F.M. (1986), 'Theories of contemporary state formation in Africa: a reassessment', *Journal of Modern African Studies*, vol. 24, pp. 335–347.

Stock, R. (1995), *Africa South of the Sahara: a geographical interpretation*, Guildford, New York.

Swan, M. (1987), 'Ideology in organized Indian politics, 1894-1948', In S. Marks and S. Trapido (eds), *The politics of race, class and nationalism in twentieth century South Africa*, Longman, London, pp. 182–208.

Terreblanche, S. and Natrass, N. (1990), 'A periodization of the political economy from 1910', In N. Natrass and E. Ardington (eds) *The political economy of South Africa*, Oxford University Press, Cape Town, pp. 6–23.

Turok, I. (1994), 'Urban planning in the transition from apartheid: part 2 – towards reconstruction', *Town Planning Review*, vol. 65, pp. 355-74.

Tutu, D. (1995), *The rainbow people of God*, Bantam Books, London.

Union of South Africa. (1959), *Hansard*, Pretoria.

Van Warmelo, N.J. (1935), *A preliminary survey of the bantu tribes of South Africa*, Government Printer, Pretoria.

Vosloo, W.B. et al. (1974), *Local government in Southern Africa*, Academia, Pretoria.

Wallace-Bruce, N.L. (1985), 'Africa and International Law – the emergence of statehood', *Journal of Modern African Studies*, vol. 23, pp. 575–602.

Walshe, P. (1970), *The rise of African nationalism in South Africa*, Hurst & Co, London.

Williams, G. (1996), 'Setting the agenda: A critique of the World Bank's rural restructuring programme for South Africa', *Journal of Southern African Studies*, vol. 22, pp. 139–66.

Williams, G., Ewert, J., Hamann, J. and Vink, N. (1998), 'Liberalizing markets and reforming land in South Africa', *Journal of Contemporary African Studies*, Vol. 16, pp. 65–94.

Young, C. (1994), *The African Colonial State in Comparative Perspective*, Yale University Press, New Haven.

Printed and bound by CPI Group (UK) Ltd, Croydon, CR0 4YY

22/10/2024

01777626-0012